TRAVELING DIFFERENT

TRAVELING DIFFERENT

Vacation Strategies for Parents of the Anxious, the Inflexible, and the Neurodiverse

DAWN M. BARCLAY

ROWMAN & LITTLEFIELD
Lanham • Boulder • New York • London

Published by Rowman & Littlefield
An imprint of The Rowman & Littlefield Publishing Group, Inc.
4501 Forbes Boulevard, Suite 200, Lanham, Maryland 20706
www.rowman.com

86-90 Paul Street, London EC2A 4NE, United Kingdom

British Library Cataloguing in Publication Information Available

Library of Congress Cataloging-in-Publication Data Available

ISBN 978-1-5381-6866-0 (cloth)
ISBN 978-1-5381-6867-7 (electronic)

For J.J. and J.J.
You've brought me my greatest challenges and my deepest joy.
I adore you both, forever and a day.

DISCLAIMER

The ability to travel is forever changing due to world, regional, and local events. What's open one day may be closed the next, and certifying organizations can alter designations of properties and venues from year to year. Special needs programs can be added, changed, or discontinued without notice. Be sure to call ahead before visiting any location mentioned herein to verify its status, exact address, phone number, hours of operation, and offerings, especially for those on the spectrum.

Please note that when phone numbers are included, they are usually direct numbers rather than toll-free numbers for the benefit of readers outside the United States. The website links that are also included will lead interested parties to toll-free numbers if they are available.

Also please note that because new autism-friendly hotels and venues open all the time, listings contained in this guide are non-comprehensive. Couple the use of this book with your own independent research. Likewise, web addresses and property addresses/phone numbers can change without notice. Feel free to send your findings and corrections to the author at dawnbarclayauthor@gmail.com.

Inclusion of venues and services in this guide does not imply endorsement. The advice offered by various organizations, parents, and travel professionals is their own.

Neither the author nor the publisher may be held responsible for errors contained herein.

CONTENTS

Acknowledgments ix

Preface xi

Part I: INTRODUCTION

1   Every Family Deserves to Travel *3*

2   Traveling with Any Child Can Be Challenging *7*

3   The Growing Focus on Traveling with Children on and off the Spectrum *13*

4   Modifications Can Help Every Family's Travel *21*

Part II: SETTING THE STAGE

5   Starting Small *29*

6   Pre-Trip Preparation *41*

7   Foreign versus Domestic? *53*

Part III: MODES OF TRAVEL

8   Navigating Airline Travel *65*

9   Road Trips—Negotiating Car and Bus Travel *89*

10   Chugging Along—Train Travel *99*

11   Smooth Sailing—Cruise Travel *107*

Part IV: ONCE YOU'VE ARRIVED

12 Staying the Course—Various Accommodations Options   123

13 The Great Outdoors—Camping, RVs, and More   145

14 To See or Not to See—Tours and Venues   157

15 Taking the Bite Out of Restaurant Dining   193

16 Other Suggested Modifications   203

**Part V: DESTINATIONS AND
SPECIAL INTEREST TRAVEL**

17 Suggested Itineraries   225

Part VI: RESOURCES AND ASSISTANCE

18 Resources   275

Notes   293

Bibliography   305

Index   315

About the Author   331

ACKNOWLEDGMENTS

This book wouldn't have been possible without the encouragement of Dr. Ellen Littman, who educated me on the ins and outs of special needs and urged me to revisit my writing career. Along with her, I must thank my agent, Rita Rosenkranz, and my acquiring editor, Suzanne Staszak-Silva, both of whom believed in me and the importance of this topic.

I owe a huge debt of gratitude to every Certified Autism Travel Professional, every parent of a child on the spectrum, and every mental health professional and advocate who shared their experience and knowledge to make this book a reality. As well as to the team at Rowman & Littlefield, whose expertise brought it to market.

To every parent who buys this book and experiments with the strategies contained herein, I applaud you! I pray it opens up the world for you and your family.

Finally, to my family, both in New York and Florida, for their unflagging support, even in my darkest moments. But especially to my husband, Joshua Jedwab, who understands that great things can happen when you are willing to endure a messy house and an empty fridge.

PREFACE

Travel has always been in my blood. As the daughter of the owners of one of Manhattan's leading travel agencies, is it any wonder? Couple that with the fact my father's entire family lived in the United Kingdom, and as you can imagine, I spent a lot of my childhood on the road.

So it was probably no surprise that when I left Tufts University, I worked at Barclay Travel Ltd., and later, Barclay International Group (a pioneer in the short-term rentals market) before my love of writing prompted me to join the editorial staff of *Travel Agent Magazine*. I traded familiarization trips for press trips as I covered "the business" of travel, attending industry conferences while often cajoling editors of other beats into letting me cover cruises and destination stories as well. Even after I got married and launched Dawn Barclay Ink—creating brochures, press releases, and other communications for, you guessed it, the travel industry—I continued to freelance for various publications and still visited exotic world locales.

That privileged life came to a screeching halt when I gave birth to two "challenging" children, about three years apart. Still, I was determined not to let their difficulties stand in my way of introducing them to the world of travel I loved so deeply. Believing there's a book to handle any crisis in life, I searched for a guide to help me tackle issues like children bawling uncontrollably as soon as they entered Galleries Lafayette, refusing to stay in their highchair at *any* restaurant, and vomiting all over my silk blouse right before takeoff when returning from the Caribbean.

In the end, we did work out our own solutions and ultimately survived, though it didn't hurt that the "challenging" kids finally grew into less-challenging young adults. But it does explain how I came up with the concept for this book.

I started soliciting quotes back in 2009, but the prospect of finding parents willing to discuss how they resolved such obstacles remained elusive. Only ten years later did I discover that the International Board of Credentialing and Continuing Education Standards (IBCCES) had created the Certified Autism Travel Professional™ training and designation. Suddenly, the pool of potential interviewees—namely travel advisors and their clients—beckoned, and there was no turning back.

This book was only possible because of those travel advisors and the parents with children on the spectrum who poured out their hearts and revealed their secret travel sauce during the hours I spent interviewing them. I owe a similar debt of gratitude to the leaders of associations and organizations like IBCCES, TravelAbility, and the Family Travel Association, who provided additional information, and mental health professionals such as Dr. Ellen Littman, Dr. Tony Attwood, and Janeen Herskovitz, who graciously shared their expertise.

It would be impossible, in one edition, to include every aspect of the hospitality industry catering to those on the spectrum, especially as new companies open their doors. If there's a travel vendor or service I've missed, please contact me at dawnbarclayauthor@gmail.com. I hope to publish regular updates at www.travelingdifferent.com.

My greatest surprise in researching this book was discovering that the tips families on the autistic spectrum use to travel the world often work for neurotypical (non-autistic) families as well. I hope this book can help soothe the hiccups of *any* family's travel. My greatest joy would be receiving a note from parents who previously thought they couldn't travel and discovered, thanks to the information contained in this guide, that the world was now open to them, and that as a family, they could build a lifetime of happy memories together.

Wishing you happy travels!

– Dawn M. Barclay

Part I

INTRODUCTION

1

EVERY FAMILY DESERVES
TO TRAVEL

"Travel and kids. Two uncontrollable variables thrown into the same mix. Some new parents are so terrified of the whole ordeal that they just put travel on hold for the first few years of parenthood."[1] – Juliet Perrachon, co-founder of Bébé Voyage, an online community of traveling parents.

"It was never easy going anywhere with my kids, but our flight back from Jamaica was a particular highlight. My one-year-old son vomited all over my silk blouse as soon as we got strapped into our seats and then burst out crying. I couldn't abandon him to change in the bathroom, so I just threw my cardigan over my blouse and was unsurprised when the passenger next to me asked the attendant if she could switch seats. Then, once the flight took off, my four-year-old daughter started screaming from the pressure in her ears. Thank God for my husband who pulled out some crayons, transformed our motion sickness bags into cartoon characters and thus, Barf Bag Theater was born. He repeated a similar feat of distraction when we tried to feed our 'spirited' children at the Rainforest Café one Sunday afternoon. He bought a toy shark puppet and proceeded to grab the children playfully with its mouth while chanting, 'Hello. I'm a shark. I won't hurt you . . . nom, nom, nom. Oops!' That's how we learned to travel the world with our kids, by using humor to rescue us during times of stress." – Dawn M. Barclay, parent and author

Anyone who's traveled with children knows it's no picnic. Be they normally easygoing or demanding, keeping kids engaged and safe can be a monumental task, especially in unfamiliar surroundings. Yet every family deserves to travel, and that is the purpose of this book: to offer tips that open the world to every parent and child, no matter what their challenges.

This book was originally meant to be a parents' guide for vacationing with children with autism spectrum disorder and co-occurring conditions such as attention-deficit/hyperactivity disorder, Tourette syndrome, borderline personality disorder, and bipolar disorder. Then my research revealed that families on the spectrum weren't the only ones apprehensive of hitting the road; interviews with travel professionals, mental health clinicians, and parents revealed that most children—even neurotypical (non-autistic) ones—become anxious when traveling outside of their comfort zone.

That revelation made writing this book even more of a priority. Family travel is a booming segment of the industry. A 2017 study[2] found that one hundred million Americans (four out of ten) planned to vacation with their families that year, and seven out of ten parents indicated they were "likely" or "very likely" to travel with their children over the next three years.[3] And even though the COVID-19 crisis played havoc with those intentions, a later poll revealed that 65 percent of Americans were looking forward to traveling with their families once restrictions were lifted.[4]

The Family Travel Association, in its 2017 survey,[5] identified three distinct groups of family travelers. The first, which the association called "Hassle-Free Travelers," prefer to stay home for vacation and avoid potential strife, or choose all-inclusive holidays, cruises, and organized tours when they do leave the house. Those in the second group, called "Cautious Travelers," might venture further afield, but due to worry over safety and keeping kids occupied, they often opt for theme park travel, "a proven draw for the younger crowd." The third group, which the survey called "Intrepid Travelers," are more adventurous, more likely to go abroad, and take their kids out of school to do so.

What I suggest they missed was a fourth group: a combination of the "Hassle-Free Traveler" and "Cautious Traveler" categories comprised of travelers with youngsters who may be anxious, inflexible, or who have been diagnosed with autism spectrum, attention, and/or mood disorders. Because the Centers for Disease Control and Prevention reports that one out of every forty-four children in the United States lives with some form of autism,[6] and the U.S. Department of Health and Human Services' Health Resources and Services Administration estimates that almost 13 percent of all families have a child with special health care needs (about 9.4 million children in total),[7] it is clearly an important segment to address.

And let's not overlook adults. The first U.S. study of autism in adults, conducted by the U.S. Centers for Disease Control and Prevention, found that 2.2 percent of American adults fall on the spectrum, a total of 5.4 million adults age eighteen and over, or about one out of every forty-five people.[8] As adults are more adept at self-control, this book concentrates mostly on children, though there is a section on traveling with older children in chapter 16.

Many parents with children on the spectrum have written books based on their experiences and insights—how they have conquered special needs travel despite its inherent difficulties—and certainly the more information out there, the better. But no books, as of the writing of this guide, have taken the lessons learned from parents of neurodiverse (spectrum) children and suggested they could also apply to issues experienced by families with challenging but neurotypical ones. And that is the mission herein: to offer parents of special needs travelers some proven travel tweaks and then explain how those modifications can simplify and enrich every family's vacation.

"Travel is one of the best gifts we can give our children, but so is teaching them that a break from the normal routine can be fun," writes Pilar Clark in TravelMamas.com:

> I'm proud of my son's atypical way of viewing the world. I encourage him to take ownership of his challenges while pushing his own limits in positive ways. Traveling with kids on the autism spectrum presents wonderful opportunities to increase awareness, acceptance, respect, and support. My family looks at it as a two-way street. Not only does our son learn new ways to experience the world, but also the world learns new ways to experience him as a person living with autism.[9]

Despite the opportunities, apprehensions exist. Take Michael Sokol, whose son Nate was born prematurely at twenty-five weeks and later diagnosed with autism spectrum disorder. The entire family started traveling to Cape Cod when Nate was around eighteen months old, but Michael admits he had myriad worries venturing further afield:

> How would Nate respond to flying? How would we navigate waiting in long lines at Disney World? Would we find food that Nate, a very picky eater, could eat? How would people react if he had a meltdown, and how would his mom handle that tantrum? Could we keep Nate safe when we were out of the country? What if he needed medical attention? Would Nate enjoy the trip? What if our flight was delayed or cancelled? Would there be Wi-Fi for Nate's devices?

Michael Sokol's fears were not atypical, which is why, in the following chapters, I will review the trials inherent in each type of travel situation, from something as mundane as packing for the upcoming voyage to tackling long bus rides, managing unexpected delays on the tarmac, sampling foreign foods, and much more. I'll offer suggestions from travel advisors, mental health experts, and parents on how to acclimate children to their new surroundings in all lodging types, how to keep them engaged, and the best ways to minimize meltdowns. Some solutions call for a change in a child's attitude and behavior, but many involve parents reframing their own mindset to embrace the essence of the family vacation—their child's experience.

"Don't insist that your children act like they're neurotypical: they're not, they can't, and why would you want them to? Those with neurodivergent brains can envision novel circumstances with unsurpassed creativity, passion, intuition, humor, and kindness," says Dr. Ellen Littman, a New York–based clinical psychologist specializing in attention-deficit/hyperactivity disorder. She suggests that the most successful family vacations occur when parents put their children's needs front and center—regardless of diagnosis or lack of same. "Avoid friction by prioritizing their interests, preferences, and sensitivities. Get a babysitter when you yearn for a high-end restaurant or a night at the theater. When parents demonstrate attuned expectations, children feel more valued, more respected, and more invested in creating treasured family memories."

This book divides travel into logical categories: different modes of transportation, destinations such as foreign versus domestic, accommodation choices, and various activities such as venues to visit. Because readers may not read cover to cover but will more likely concentrate on whatever transportation mode and activities their upcoming vacation involves, you might find some repetition in these pages. There is also some cross-referencing of chapters, so you don't miss any important advice. The book will culminate with a list of destinations and itineraries specially attuned to the needs of travelers on the spectrum, but which cater to all families.

So, sit back and enjoy the journey, for it is designed to enrich all the journeys that follow.

2

TRAVELING WITH ANY CHILD CAN BE CHALLENGING

"Children can be grumpy, uncooperative, argumentative, or tired. They can easily spoil your plans and be their own adorable little travel horror story. It's best not to get too hung up on plans. Occasionally they make our lives more difficult, but they're beautiful, fun, amazing little human beings and I wouldn't have our lives any other way."[1]
–Alyson Long, Worldtravelfamily.com

Because all children—not just those on the autism spectrum—are creatures of routine, any parent with travel ambitions can immediately sense where difficulties may arise:

- How will changes in their schedule affect play and interaction?
- What new foods or environments might stimulate allergies?
- What new foods and flavors will children accept or reject?
- How will children react to the overstimulation caused when encountering unknown places, such as airports and tourist attractions?
- How will children adjust to new sleeping conditions?
- During the trip, will children be able to communicate what's really bothering them?

Picture the daunting prospect of traveling with children who haven't eaten or slept. Or who can't tolerate crowds in department stores, theme parks, or during a cruise ship's lifesaving drill. One can easily imagine that, without adequate preparation, the vacation might not end well.

Some challenges experienced by children on the autism spectrum also affect neurotypical children, but to a lesser degree. Consider these characteristics of autism spectrum disorder (ASD) as described by the American Psychiatric Association: "Reduced sharing of interests or emotions; challenges in understanding or responding to social cues such as eye contact and facial expressions; having a significant need for a predictable routine or structure; and deficits in developing/maintaining/understanding relationships (trouble making friends)."[2] What parent of a neurotypical child hasn't struggled at some point to extract a modicum of excitement from a sullen toddler, corrected wavering eye contact when addressing their son or daughter, dealt with a meltdown over an unexpected scheduling change, or intervened to settle an argument between squabbling playmates? Is there any doubt that the tweaks for the ASD group might benefit travel for all?

"Travel is an adventure," says Dr. Ellen Littman, a New York–based clinical psychologist who works with children with attention-deficit and hyperactivity disorder (ADHD) and ASD. "What this means is that no one can predict or control what happens when they're traveling. [By communicating this,] children can then understand that even parents may not be pleased by everything [they encounter]. So. it's far less likely that there will be whining when [children] understand that no one in the family knows what's around the next corner; everyone is on an adventure of exploration and discovery."

The key to success for both populations is to "allow your child ample time for transitions," says Angela Romans of Rejuvia Travel in South Bend, Indiana. "As often as possible, try to keep a routine but prepare for the effects of a disrupted [one]. Reduce your own stress as much as possible. Try to understand that new environments and situations can be overwhelming. Be patient and put yourself in their shoes."

Of course, not all children with autism exhibit the same behaviors, or to the same degree. ASD is precisely that: a spectrum. Some autistic characteristics diverge more dramatically from the neurotypical, such as "restricted and repetitive patterns of behaviors, interests and activities; playing with toys uncommonly; speaking in a unique way; exhibiting intense interests in activities that [vary] from a similarly aged child; and experiencing the sensory aspects of the world in an unusual or extreme way."[3] Therefore, this book offers suggestions for both high- and low-functioning individuals, even when they won't necessarily apply to neurotypical families.

Lisa Bertuccio of Vacation Your World in Otisville, New York, advises: "The use of autism awareness cards can be beneficial to have on hand during travel. Parents can hand these cards to onlookers to explain why their child may be behaving atypically. This helps to avoid confrontation and adds awareness so that in the future, onlookers can be more understanding."

Another pervasive special needs travel category covered in this book is ADHD. While not on the spectrum, the Centers for Disease Control and Prevention report that almost 10 percent of children aged two to seventeen have been diagnosed with ADHD,[4] and the advocacy group CHADD estimates the disorder affects 129 million children worldwide.[5] Symptoms include difficulties with "executive functioning skills, . . . attention, concentration, memory, motivation, and effort, learning from mistakes, impulsivity, hyperactivity, organization, and social skills,"[6] which are also behaviors frequently seen, if to a lesser degree, in the neurotypical community.

Michelle Heubish of Smithtown, New York, travels with three children, one of them diagnosed with ADHD. She considers how tweaks like advance research, frequently advised by family and autism travel professionals, might have prevented this unfortunate occurrence at a theme park:

> What's hard with an ADHD child is that they are not very "go with the flow" and don't get over things and move on easily. My son perseverates, rehashes things over and over. . . . When we went to Disney World for a week, we did all four parks, character meals, you name it. He does not like rides that go upside down. Well, my sister told him that Rock n' Roller Coaster didn't go upside down. I had never been on it, and she hadn't been on it in twenty years—so we go on it [and learn] it was a ninety-second ride [that] went upside down. He was upside down for maybe ten seconds. A year later, he won't tell you about the eighty rides he loved, or how he had lunch with Mickey, or any of that. He will just tell you how we "lied to him" and he hates Rock n' Roller Coaster and he's never going to Hollywood Studios again. There is no solving that—sometimes life throws you unexpected curve balls and you aren't ready. You don't get a color-coded itinerary for everything in life.

Likewise, those diagnosed with bipolar disorder—a mood disorder rather than a developmental one—may not be on the autism spectrum, but because "bipolar disorder and autism share several common symptoms and behaviors . . . some people with ASD may be mistakenly diagnosed as bipolar."[7] This mother of a bipolar daughter learned a lesson that might help all travelers: every child is different. Never assume one size fits all: "Long before 'special needs travel' was widely discussed and accepted, we'd travel to the Caribbean every year with my husband's best friend from law school and his family," says a mom from New York who requested to remain anonymous and that the names of the children be changed. She continues:

These friends were tolerant of Jaseree's inflexibility, though skeptical of her bipolar diagnosis, probably because they didn't spend much time with her—most days, we kept Jas ensconced in the kids' club. But after one trip to St. Thomas when our daughter was around ten, that all changed. I was sitting on the plane with Jas and Amy, our friend's daughter, when mid-flight, Jas had a meltdown. During a dark half-hour, she spewed venom, disparaging life, family, vacations, and especially kids' clubs. Turned out she hated kids' clubs. She even threatened suicide. I could tell the other girl was shocked, but being mature for her age, she tried to hold it together. I finally talked Jas down, but the damage was done. That other family politely begged off of traveling with us ever again, and when we next ventured out, we kept Jas by our side since she preferred to be around adults anyway. They didn't judge her as harshly as kids her own age.

Neurotypical children may experience some symptoms of bipolar behavior but again, like ADHD and ASD, to a lesser degree. According to the Depression and Bipolar Support Alliance, these include, but are not limited to, issues with irritability, problems with sleep, impulsivity, poor judgment, and a tendency to be easily distracted. These can alternate with unprovoked crying, anxiousness, concentration difficulties, lack of interest, and phantom aches and pains.[8] Doesn't that sound like every child? For those reasons, I've included tips for this population as well.

Keeping these similarities in mind, let's move on.

PARENT SPOTLIGHT—
MICHELLE HEUBISH, SMITHTOWN, NEW YORK

Michelle Heubish has three children, one diagnosed with ADHD and dyslexia. She's been traveling with them since her oldest was one, seeking relaxation and a chance to expose her children to the world, though she usually stays domestic due to the additional expense of venturing abroad. She believes that some parents with special needs children don't follow in her footsteps because they "are afraid that their children can't do what others can. The truth is, while some experiences may need to be modified or certain aspects avoided, if given the opportunity, they can do almost anything neurotypical kids can do. For example, my son does not like loud noises so we need to bring headphones for him—that is a small modification but allows him to still enjoy experiences [he might have otherwise missed]."

Adjustments may only go so far, she admits:

> My son with ADHD is the oldest so sometimes the others follow his behaviors: the complaining, the lack of patience. Recently we were at a hotel with an indoor pool—this was a big treat because we live in New York where it is cold in the winter, so they hadn't gone swimming in months. My "plan" was to let them swim for an hour or two in the morning before checkout and expend some energy before the long drive home. My younger two were happy as can be, jumping in, etc. [but] my son with ADHD became bored with swimming after about twenty minutes.
>
> As soon as he decided he was "done" and got out of the pool, the other two followed suit. They would have stayed in there another hour if it weren't for him. My younger children seem to be happy with most things—going to a farm or a zoo or an aquarium for example, but these things are not "exciting enough" for my ADHD child. He needs to keep moving constantly. Often, we haven't even finished one activity, and he is already asking about what is next. The other children seem to live "in the moment" more.
>
> I usually have to tell my ADHD child what he can expect ahead of time. I call him the "cruise director" because he will literally hop out of bed and say something like, "What's the plan for today, Mommy?" I always give my son an idea of what he can expect. We've tried surprises and those result in stress.

Heubish's advice when traveling is to hold hands and have a protocol in place should you get separated. For airline travel, bring things to keep the child busy: coloring, iPads, LEGO® sets. When booking accommodations, "extra space is always helpful. Don't try to cram into a standard hotel room with two or three kids. The additional space is worth the extra few dollars; save money elsewhere."

When touring, Heubish says she avoids organized tours. "I make my own list of must-sees. That way, [if] we need to leave, we can do so without disturbing other people." When navigating restaurants, she says she tries to

> know what they want to eat beforehand [and] order the kids' food first to eliminate [having them] wait. Another tip someone shared with me

is if you have a kitchen of some kind while traveling, feed the kids in the hotel and allow them to order dessert while the adults eat dinner. This cuts down on the cost but also the waiting—ice cream takes five minutes while a meal takes much longer.

Preferred destinations?

I don't have specific ones, but we like to mix natural attractions with commercial fun. For example, we went to Maine this summer. One day we hiked Acadia National Park, [and on subsequent days] we went to the Lumberjack Show, a Lobster Boat, the beach where we saw the lighthouse. Same thing with Niagara Falls—one day we did the Hornblower Cruise under the Falls and took the tour behind the Falls but then at night, we let them play mini-golf and ride the Ferris wheel. We try to mix it up.

A final piece of advice:

We travel a lot with our children (ages nine, six, and three) and people tell us to wait until they're older and they'll remember it more, or when it will be easier to travel. I say, they may not remember it all—but my husband and I will. Some of my favorite memories of my children are from our travels. When you get away from home, you focus on where you are and who you are with—not the laundry, not the dishes, not the homework. Parents would be surprised what their kids remember and get out of traveling.

I have had many of my kids' teachers tell me about the connections they have made in school because of our travels. In class, when they learned about the upcoming election, both of my boys remembered going to Washington, DC, and seeing the White House as well as some random facts like the president having his own bowling alley. The other thing is—and parents don't like to think about this—eventually your kids will get to their teenage years. They will have their own lives, sports, and girlfriends/boyfriends, and will not be able to (or want to) travel with you, so this is the time—enjoy it!

3

THE GROWING FOCUS ON TRAVELING WITH CHILDREN ON AND OFF THE SPECTRUM

"You can't put a price tag on seeing the bewilderment in your child's eyes the first time they see the waterfalls of Yosemite National Park, or the exhilaration that comes from Mom, Dad, and the kids putting on their masks and snorkels to swim together with tropical fish in the Caribbean. I can also vividly remember high-fiving my eight-year-old son after surviving a whitewater-rafting trip we took on the Snake River outside Jackson, Wyoming." – Rainer Jenss, founder, Family Travel Association

"It's not like all my travels are filled with memorable moments of snapping photos of butterflies and practicing a new language with my kids. I've had my share of challenges. I lost Kai at our condo in Kauai; once he stuck his finger in a moving fan while I was having a foot massage in Bangkok; another time he vomited all over me on a plane ride and the flight attendant shrugged, gave me wipes to clean it up, and then I had to sit in the stench for four hours. Trust me, the list goes on." – Michele Bigley, author of *Sometimes Travel with Kids Sucks*[1]

Two stories, two different experiences. Luckily, though, for those who want to pack up their kids along with their belongings, travel has never been so child friendly, with plenty of resources available to help ensure your vacation is more like the first scenario than the second.

A BRIEF SPECTRUM PRIMER

For travelers curious about how those on the autistic spectrum might differ from neurotypical children, I will include this brief primer—abbreviated specifically because laypersons aren't reading this book for a lesson on autism and those with children on the spectrum are already aware of the dissimilarities.

According to the National Institute for Mental Health, clinicians once broke autism down into several conditions, including autistic disorder, Asperger's syndrome, and pervasive developmental disorder not otherwise specified. Some authorities included childhood disintegrative disorder in that definition as well. "But in 2013, a revised version of the *Diagnostic and Statistical Manual of Mental Disorders (DSM)* was released. This revision changed the way autism is classified . . . into one diagnosis called 'autism spectrum disorder' [ASD]. Although the 'official' diagnosis of ASD has changed, there is nothing wrong with continuing to use terms such as Asperger's syndrome to describe oneself or to identify with a peer group."[2]

The National Autism Association defines autism as a

> bio-neurological developmental disability that generally appears before the age of three. It impacts the normal development of the brain in the areas of social interaction, communication skills, and cognitive functions. Individuals with autism typically have difficulties in verbal and nonverbal communication, social interactions, and leisure or play activities. Individuals with autism often suffer from numerous comorbid medical conditions which may include allergies, asthma, epilepsy, digestive disorders, persistent viral infections, feeding disorders, sensory integration disorders, sleeping disorders, and more.[3]

Sensory issues often accompany autism. In 2013, the American Psychiatric Association added sensory sensitivities to the symptoms that help diagnose autism. These issues can involve both hyper-sensitivities (over-responsiveness) and hypo-sensitivities (under-responsiveness) to a wide range of stimuli, including sights, sounds, smells, tastes, touch, balance, and body awareness. For example, many people on the spectrum are hyper-sensitive to bright lights or certain light wavelengths (e.g., from fluorescent lights). Many find certain sounds, smells, and tastes overwhelming. Various types of touch (light or deep) can feel extremely uncomfortable.[4] In this book, you'll see many tweaks that focus on mitigating these sensory issues.

According to *Autism Parenting Magazine*, the DSM-5

> categorizes autism into levels. Level 1 is the highest-functioning category. Children with Level 1 autism need support but can learn a variety

of social skills. They can usually gain some independence. Individuals diagnosed with Level 2 autism have verbal, social, and behavioral deficits. Even with supports in place, they might struggle with these behaviors. Level 3 is the most severe and lowest functioning category. Those with Level 3 autism have immense difficulty socializing, speaking, and communicating nonverbally. These children experience great distress in situations outside of his/her comfort zone.[5]

This differentiation is under attack by some spectrum researchers who argue that the distinction should be scrapped. They believe that people with ASD described as "high functioning" due to their lack of intellectual disabilities often still struggle with daily living skills. They point to an Australian study of over two thousand people on the spectrum, which found that individuals deemed high functioning often have poor "adaptive behavior"—the ability to perform basic tasks such as brushing teeth, tying shoelaces, or taking the bus.[6]

Interestingly, attention-deficit/hyperactivity disorder (ADHD), a diagnosis which now incorporates attention deficit disorder, is not classified as an ASD. According to the Attention Deficit Disorder Association, "ADHD is a highly genetic, brain-based syndrome that has to do with regulating a particular set of brain functions . . . collectively referred to as 'executive functioning skills' and include . . . attention, concentration, memory, motivation and effort, learning from mistakes, impulsivity, hyperactivity, organization, and social skills."[7]

However, according to Elizabeth Harstad, MD, MPH, the two are related in many ways. She says that ASD and ADHD have similar symptoms and having one might increase the chances of having the other.[8]

Consider her rationale regarding the overlap, as described by Understood.org:[9]

- Trouble paying attention: Kids with autism may struggle with this for several reasons. One is that language difficulties can make it seem like kids aren't paying attention to directions. But it may be that they just don't understand the directions.
- Trouble socially: ADHD can affect social skills. This can include avoiding eye contact and getting into other people's personal space.

While it's true that some children with ADHD may present diametrically opposite to those with ASD, both require similar travel tweaks.

Because the Centers for Disease Control and Prevention report that almost 10 percent of children aged two to seventeen have been diagnosed with ADHD,[10] and that according to Children and Adults with Attention-Deficit/Hyperactivity Disorder, 129 million children worldwide live daily with this issue,[11] they can also surely benefit from the suggestions offered in these pages.

Likewise, those diagnosed with bipolar disorder may not be on the autism spectrum, but experts say that because bipolar disorder and autism symptoms and behaviors resemble one another, those with ASD may be mistakenly diagnosed as bipolar. Sleep issues are prevalent in ASD,[12] and those with bipolar disorder can be triggered by changes in sleep patterns (i.e., jet lag,[13] a frequent component of travel). Therefore, many of the travel tips for ASD may apply to those with bipolar disorder as well.

NEUROTYPICAL FAMILIES HAVE ISSUES TOO

Thanks to an ongoing focus on family travel, it's never been easier to travel with kids. *The Family Travel Handbook* reports that

> there's a huge range of accommodation options [available], and many more sights and attractions are geared towards entertaining small people. And that's without even mentioning the benefits of technology. Digital maps, translation apps and games for the kids all make life on the road much less stressful than was previously the case.[14]

And yet, issues persist. I posted a question in a Facebook travel group asking parents about their biggest problems when traveling with their kids. The question did not differentiate between neurotypical and neurodiverse family travel. Members posted 175 responses over the next forty-eight hours, their comments involving issues with attention, eating, sleep, and medical issues.[15] Among their complaints: toddlers crying on airplanes; how restaurant tables, hotel rooms, cruise cabins, and ride shares are only configured for parties of four (and not larger families); keeping children occupied during flight delays; constant bathroom stops on road trips; alleviating boredom—especially when children miss their friends; conquering jet lag and adapting to time changes; dealing with motion sickness; finicky eating when faced with new foods; sibling disputes; handling medical emergencies away from home; and how it just wasn't the same as traveling before kids when they could go at their own speed and visit upscale restaurants.

To deal with issues like these, Rainer Jenss founded an organization called the Family Travel Association (FTA) in 2014, designed to "inspire and empower families to travel and help them discover what's possible."[16] Such possibilities include exposing kids to new cultures and perspectives, fostering self-esteem and independence, strengthening their connections to wildlife and the natural world, forming lifelong friendships, and teaching them new life skills. His family travel philosophy? "When you learn about the world, you learn about yourself."

Sadly, families with anxious, inflexible, or neurodiverse children haven't always felt like they had access to such experiences, according to travel professionals who handle travel for both populations.

"Some believe having a family member on the spectrum permanently eliminates them from family travel," says Betty J. Burger, recently retired from BGB Travel and Leisure, LLC, in St. Marys, Georgia. The media sometimes perpetuates that misperception, notes Starr Wlodarski of Cruise Planners—Unique Family Adventures in Perrysburg, Ohio: "They see reports on TV that kids with autism get kicked off planes."

This is unfortunate because traveling together can provide many benefits for special needs families, including the opportunity to reduce any self-imposed isolation while strengthening bonds within the unit. Travel also can introduce children on the spectrum to new life skills, such as flexibility and the acceptance of new cultures. Another plus: a trip away from the familiar is a way to desensitize a child with autism to challenging sensory stimulation and to spread autism awareness.[17]

And it's not for lack of interest. According to a recent International Board of Credentialing and Continuing Education Standards (IBCCES) study of families on the spectrum, 87 percent of one thousand parents interviewed said that they have never traveled with their child, yet 93 percent of that subset said they would travel if there were more autism-friendly places to go.[18]

Worry often overrides desire, however. Parents report concerns over costs, dealing with safety issues, the stigma of appearing "different," and, especially, the stress of feeling judged. What might be a reaction to overstimulation or difficulty communicating emotions could easily be misinterpreted as a tantrum, they say. "People that don't have autism in the family don't understand the behaviors. Most people think the child is a 'brat' or not disciplined," explains Brenda Revere of C2C Escapes Travel in Bedford, Indiana.

Other families on the spectrum worry about having to explain themselves and their situation, according to Michelle St. Pierre of Modern Travel Professionals in Yardley, Pennsylvania:

> Parents (and maybe to a greater degree, grandparents) believe they have to bring evidence of a diagnosis to the airport, to Disney, or on their cruise to have accommodations made for their child. They don't realize that most times travel operators are happy to provide accommodations (within reason) if you ask for them and explain why you need them, and that they legally can't ask for a diagnosis, anyway! I'm always being asked by parents or grandparents, "Is this doctor's note going to be good enough?" and I tell them to practice asking for what they need instead of stating the diagnosis of their child.

To ensure that they focus on travel for all families, the FTA has teamed up with the IBCCES to ensure that travel agents and suppliers all over the globe have access to training and certification in the special needs market. The result: the Certified Autism Travel Professional™ (CATP) program for travel advisors and the Certified Autism Center™ designation for travel vendors. Considering that ASD is the fastest-growing serious developmental disability in the United States,[19] and, according to IBCCES, grows at a 600 percent increase year over year, the time is ripe to "provide travel solutions for a $262 billion underserved market."[20]

The comprehensive CATP program covers ten competencies: What Is Autism?, Selling into the Autism Market, Breakdown of the Travel Market, Sensory Awareness, Individual's Perspective, Parent's Perspective, Resorts/ Hotels, Air Travel, Cruise, and Road Trips. Continuing education is required to renew the designation, guaranteeing that CATPs are kept aware of the latest industry developments.

Michelle St. Pierre says she pursued the CATP designation because

> I wanted to be very open with families about the assistance I could offer. . . . I found some parents danced around their issues or were hesitant to ask questions about how travel might be experienced differently by their children on the spectrum. By getting the CATP designation, I could "put it out there" that I could help with the questions and issues they faced, rather than leaving them at the mercy of searching for answers on the internet.

"I have spoken with other parents of special needs children who [don't believe] their child will benefit from travel," says Michael Sokol, whose son has ASD. He continues:

> Many kids with autism do not speak or speak a little, but just because they don't verbalize about a vacation doesn't mean that they didn't enjoy it. Traveling opens the mind. Seeing and experiencing new and different places has been great for Nate's development. A well-versed travel agent who understands special needs can provide a huge advantage in arranging a seamless vacation.

Many CATPs direct their clients to Certified Autism Centers. To become autism-certified, a venue (hotel, resort, cruise line, attraction, restaurant, etc.) must dedicate itself to "serving individuals with autism, have at least eighty percent of its staff trained and certified in the field of autism, and be committed to ongoing training in autism."[21] You can read more about their certification process in chapter 5.

The Center for Autism and Related Disorders (CARD), which calls itself the world's largest treatment provider, also certifies properties as spectrum-friendly, and that's something Tara Woodbury of Escape into Travel in Leland, North Carolina, looks for when advising her clients. When a property is CARD certified, she says,

> it means the staff receives annual training on providing a better experience to guests and families with special needs. Some hotels offer "safe kits" for the room, including things like door alarms, corner cushions, etc. If you need something like safety rails for the bed, ask if the hotel can provide them. If not, there are rental companies who can provide such items. . . . CARD-certified hotels may also offer downloadable social books to help with your stay. Some, like the Tradewinds Resort in St. Petersburg, FL, offer kids' clubs with autism and special needs–friendly activities. And hotels that offer room service or to-go meals (meals that don't require cooking) might be a good fit for days when your child has had enough time around other people.[22]

While these are wonderful developments, it would be a mistake for parents or suppliers to believe that when making special needs modifications, one size fits all, warns Sarah Marshall of TravelAble in Midlothian, Virginia. "Every child on the spectrum is different. They are going to have different needs and requirements." This is something to consider as you read this guide: while one piece of advice may work for your child, on the spectrum or off, another might not. It's a matter of trial and error, made easier by learning from the successes of others.

ADVOCATE SPOTLIGHT: RAINER JENSS
ON TRAVEL FOR SPECIAL NEEDS FAMILIES

It's estimated that over 12 percent of families has a special needs child,[23] and Rainer Jenss, the founder of the FTA, "has given the floor" to organizations that cater to these families. For example, "one of our members, Il Viaggio, is a ground operator that's taken the initiative to make Costa Rica a destination that's more friendly and accessible to all families. They've set up the first beach that caters to this audience, including the ability to bring wheelchairs and other special needs equipment" closer to the water's edge. He also commends Royal Caribbean International for extending special needs protocols to Celebrity Cruises.

"Vacation rentals are another excellent option for parents with children on the spectrum and are here to stay," says Jenss. "VRBO, which now owns HomeAway, has worked to make homestays safer and hosts easier to deal with." He adds that many vacation rental companies like Wyndham have professional management, ensuring a consistency of service.

When planning successful family trips—on or off the spectrum—Jenss says the key is "to get the kids involved in the planning. It will give them a vested interest in the trip because they were part of it, which means they are much more likely to enjoy it." He advises that you can empower children by letting them make suggestions, but adds, "Don't feel you have to abide by everything they say."

Another essential tip Jenss offers to all family travelers: First choose an activity or type of experience you seek and *then* select your destination. Otherwise, you're doing it backwards and may limit yourself in what you can do based on what that destination offers. "Let's say you want an active vacation with a beach, a kids' club, and pina coladas. But you settle on the destination first. What if that destination turns out to be too expensive? Or if your chosen dates coincide with their rainy season? You've locked yourself in because you chose the destination first. But if you planned the other way, you've just got a lot more to choose from. It's important to have options," he says.

4

MODIFICATIONS CAN HELP
EVERY FAMILY'S TRAVEL

*"Things like visual schedules can help **all** kids feel secure because they know what to expect."* – Kristen Chambliss, parent, League City, Texas

While researching this book, I was struck by how many of the families I spoke with had children both on and off the spectrum within the same household. Several told me how the modifications they'd made for one child often helped the other. The truth is that children are creatures of habit, and travel is a disruption of everyday life. For that reason, even neurotypical children may need some hand-holding when venturing far from home.[1]

Therefore, parents of neurotypical children shouldn't be afraid to try some of the travel tweaks mentioned in this book. Interestingly, in Carol Arky's article for the Child Mind Institute titled "Tips for Traveling with Challenging Children,"[2] much of the advice directed toward neurotypical, albeit anxious, children mirrors the advice offered by special needs parents in these pages.

The key is to not be put off by the term "special needs" and to be open to whatever might soothe the hiccups of travel for all family members. One bonus: implementing modifications for everyone might make the anxious, inflexible, or neurotypical child feel less "different" and alone.

While some travel tips and accommodations are specifically for families with someone on the spectrum, many are more general in nature and can absolutely alleviate anxiety in neurotypical families, says Jennifer Hardy of Cruise Planners in Kent, Washington:

> Imagine being a young child going to the airport for the first time and having to wait in a long line for security, seeing adults around you

21

acting oddly, people unpacking their bags and removing articles of clothing before they and their belongings go through strange-looking machines. All that, combined with terrible lighting, lots of people, and noise would create anxiety and a potential for meltdowns in any child. Preparation tips such as utilizing Social Stories™, watching YouTube videos, and reading books about the airport experience will help a child understand this is normal, expected, and safe. (Read more about Social Stories in chapter 6.)

One tool, repeated many times throughout this book for those who aren't reading cover to cover, is the use of the "magic" or "go-to" bag which, says New York–based clinical psychologist Dr. Ellen Littman, when properly packed and managed, can save relationships. "Only a parent carries this bag and can use one item whenever boredom, overwhelm, or fatigue threatens to rock the boat," she explains. Her suggested inclusions:

- Noise-canceling headphones, to reduce stimulation and sensory overload
- Snack-size Ziploc bags with low/no sugar foods in discrete units and disposable packaging that are not messy, don't need refrigeration or utensils, like popcorn, cereal (Cheerios, Kix, etc.), cut-up carrots, animal crackers, and juice boxes. Sugar will jazz them up and make it harder to sit still.
- Surprises, costing less than ten dollars, in gift bags. Think Silly Putty, four-color pens, etc. Nothing that makes noise or moves and might annoy others.
- A change of clothes, so you can be prepared rather than frantic when things get wet, dirty, etc.
- Art supplies or Etch-a-Sketch
- Lots of Band-Aids, Neosporin (and medication as approved by your own physician), Benadryl, anti-nausea medication, Tylenol

Also, "allowing time for breaks in your schedule is important for all families," adds Jennifer Hardy. "Every child needs a quiet place to rest, especially if they are walking for most of the day. Scheduling a time and place for that break will allow the entire family to last longer and experience more."

Another of Hardy's tips is pre-planning an itinerary in detail, making sure to enlist the input and assistance of any children traveling along. She explains:

This sets the expectation of what the experience will be like, including an understanding of times when there will be excitement and times when there will be waiting. This can also give you a preview of their reaction to certain activities—something they love, and you want to spend more time doing, or something they detest, and you want to reallocate your time elsewhere. Finally, including your children in the planning process teaches them valuable skills such as preparation, goal setting, and creating backup plans. This allows your children to become expert-level world travelers in the future!

"My family and I went to Florida a few months ago," says Lisa Bertuccio of Vacation Your World in Otisville, New York:

It was our younger boy's first time flying. I found myself making sure I explained what to expect at the airport, on the flight, and once we landed. I scheduled activities during our vacation around my children's typical daily schedule. [For theme parks], I had to show one of my sons some videos of the ride he would be going on [in advance] because he would be less afraid if he knew ahead of time what the ride would be like. I also took precautions in case one of the kids got lost by taking their picture each day in the current clothes they were wearing. All these actions are tips I'd suggest for all families regardless of whether they have a family member on the autism spectrum or not.

"Often, the only difference in implementing these tips is the way they are carried out," says Bertuccio:

Having a schedule is important for all kids. Neurotypical children may be fine with a simple explanation of the schedule for the day whereas children on the autism spectrum may benefit from a visual schedule [where a string of pictures illustrates a sequence of planned activities or the steps necessary to complete a specific activity].[3] We may prepare for and combat a meltdown with neurotypical children by making sure we stop and rest somewhere or sit down for a small snack, but with children on the spectrum, the preparation and remedy for a meltdown may mean finding a quiet place or playing with a manipulative [a toy that encourages mathematical skills, counting, sorting, and learning patterns, often tactile and sensory].[4]

"Kids are kids, no matter what," she adds. "They will have meltdowns. No one child's preferences or tolerance levels are the same as any other child's, and their attention spans vary as well."

Meredith Tekin, the president of International Board of Credentialing and Continuing Education Standards—which trains and designates properties as Certified Autism Centers—makes it clear that

> at the end of the day, being more empathetic, understanding, and knowledgeable about people's various needs not only helps those individuals with those specific needs, but also opens up professionals to providing better customer service overall. In addition, a lot of the recommendations, sensory guides, and information we share with our certified partners can help parents or individuals with a variety of needs, those who want more preplanning details, or those that just want a better understanding of the experience they will have once on site. We also ensure we include not only a variety of perspectives in our programs, but also [a directive] to treat everyone as an individual first, which again, helps all guests.

One challenge, however, exists in families who have both neurodiverse and neurotypical children. Dr. Tony Attwood, a noted autism spectrum disorder (ASD) expert, suggests parents pay special attention to the child or children off the spectrum:

> Having a happy child with autism means having a happy family on vacation. There can be issues of the siblings resenting the child with autism, from shortening trips to interesting vacation destinations to taking up too much of parents' time. Neurotypical siblings need to be part of the solution, not the problem, and will need to be included in designing the strategies of how to cope with a brother or sister with ASD on vacation and while travelling.

PARENT SPOTLIGHT—ADVENTURE-IN-THE-MAKEING, FROM NOVI, MICHIGAN

A mother who wishes to be referred to as "Adventure-in-the-Makeing" (sic) has two children, one on the spectrum with ASD, attention-deficit/hyperactivity disorder, and anxiety issues. She says:

> We started traveling with him when he was six months old, eager to relax and expose both kids to the world, despite the apprehension that it would be difficult to maintain routine. What we learned is that you

can never prepare enough. Onlookers have the misconception that kids with ASD can "act normal" if they want to; because my son is high functioning, people think that when he mimics others, he's just being bad.

We think ahead when we pack, considering unforeseen stops, and we make sure to have medication and other tools easily accessible. This includes iPads and headphones, which we bring wherever we go—except restaurants, that's family time—along with a generator so we can recharge them in the car. We show our son YouTube videos in advance, so he knows what to expect, but we emphasize that rules remain the same, they don't change just because we're on vacation. We have accessibility passes for venues that are autism certified. And we never make the kids do stuff just because we want to do it, or because we've already paid for it.

Our advice to other families on the spectrum: Don't stop traveling because it's hard, or it takes more preparation. My son has done things we never did, and he will be a better adult because of it.

CHAPTER 4 ACTION STEPS

1. Don't be afraid to use tips in this book, even if your child is neurotypical. Put yourself in their shoes, and imagine how disruptive travel can be to the routine they cling to.
2. Include breaks in your schedule for downtime.
3. Pre-plan and explain as much as possible to your child in advance of the trip.
4. Schedules are your friend. They give children a roadmap of each day's activities. Use visual schedules if that's what works best.
5. If you have both neurodiverse and neurotypical children, make sure you pay equal attention to both to ward off sibling resentment.
6. Keep medication and sensory tools easily accessible, especially in transit.

Part II

SETTING THE STAGE

5

STARTING SMALL

"Start with a small trip, close to home. Increase your distance as your child grows comfortable. You don't want to be far from home on your first trip and be unable to return if things don't work out as planned."
– Jennifer Hardy of Cruise Planners, Kent, Washington

They say that the best way to eat an elephant is one bite at a time. When introducing anxious children to the idea of stepping outside their home—and their comfort zone—parents who've successfully overcome this obstacle suggest taking many small steps that can ultimately lead up to the big one: a full-blown vacation.

Michelle Zeihr, a Canadian from Barrie, Ontario, is one such parent, and her advice is to start kids traveling as young as possible. That way, it comes naturally to them. She and her husband have two children, Thessaly, a nine-year-old daughter with high-functioning autism spectrum disorder (ASD) and attention-deficit/hyperactivity disorder (ADHD), and Eden, age four, who appears neurotypical but is under observation by an early intervention team, "just in case."

Michelle and her husband started with small trips in their province when Thessaly was only five months old, introducing her to tent camping and a cabin stay. At nine months, they ventured further afield, taking a road trip across the border to Buffalo. "We wanted to see how she'd do in a hotel," Zeihr says. At fourteen months, they took her on her first plane ride and "real" trip to Mexico. "We've traveled almost every year with her since, except 2020–2021, due to the pandemic."

"Start with trips to the grocery store. Then day trips. An overnight to a local hotel. A two-day road trip somewhere close by. A three-day stay at a theme park. A week-long trip that includes an airplane ride and new experiences," advises Nicole Thibault of Magic Storybook Vacations in

Fairport, New York, laying out a blueprint that can work with neurotypical children as well as those on the spectrum.[1]

Taking short trips using your own house or apartment as home base (i.e., a staycation) helps save on transportation and housing costs when you're experimenting with travel on a smaller scale. It's also an opportunity to look at every local venue as a learning opportunity, says G.G., a mother from New York who prefers to remain anonymous. "A local special needs parents group organized a visit to a neighborhood supermarket when my children were small. There were about five families involved and we toured behind-the-scenes, including the butcher's station and the freezer areas. The kids were fascinated, and it was a good, short introduction to what an organized tour would be like."

The zoo is a perfect outing for kids with autism, says Thibault. "There is a low chance for sensory overload, and you can walk around the exhibits at your own pace. You can bring your own snacks and drinks for those kids on special diets or with a limited food repertoire. And you can avoid areas that may be too overwhelming for your child.

"Animals can also be a [great] introduction [to travel for] kids who love 'fun facts.' We'd spend time exploring the exhibits and learning the names of all the animals. Then my sons and I would head to the library and check out books about those animals of interest," she adds.[2]

Parents suggest that introducing travel on a local level can be as simple as reading timetables at a nearby train station or bus stop, visiting a nearby museum or farm, reframing a day of garage sale shopping as a "treasure or scavenger hunt," apple picking at the neighborhood orchard, taking a hike or bike ride together, enjoying a picnic lunch, or even setting up a tent and sleeping bags in the backyard. It's a way to slowly bring new experiences into children's lives without disrupting their entire routine.

Next, you might schedule an overnight at the home of a close and empathetic family member or friend. Experiencing the novelty of an unfamiliar sleeping space for a short stay can usher in the possibility of longer stays at hotels or vacation rentals.

"To introduce the idea of foreign travel, each month, we would read picture books about a different destination and then study various aspects of that country," says G.G.:

> March might be all about Italy. I'd make pasta and pizza for dinner and teach the kids some easy Italian words. We'd watch a travelogue about

Italy on the internet, maybe talk about what their money looks like, discuss some small cultural differences. It was about teaching them that there are people all over the world living lives different from theirs, and maybe sparking some interest into visiting those countries one day in the future.

Those who live in cities teeming with diversity might augment "country studies" like G.G.'s by seeking out local ethnic bakeries, restaurants, and shops selling food and wares reflecting cultures different from their own. Such a field trip might provide a child with a glimpse into foreign travel without leaving town. The children's picture book *Madlenka* by Peter Sis describes such an adventure. Perhaps reading about Madlenka's escapades could inspire an anxious child to duplicate her quest. I've listed more such books later in this chapter.

Indulging a child's passion—be it for trains, rocks, or whatever their fancy—might be a great way to ease the discomfort of a change in schedule. Be creative, and do some research. Taking short train trips might smooth the way for a railfan to endure the restaurant stop that follows. You might satisfy your rock-loving child's curiosity by visiting a local quarry or a store selling granite, marble, and tile for kitchen/bathroom renovations, or perhaps grab a shovel and pail and dig up a small corner of your backyard. You could even embark on a virtual geology field trip through a website like Micromyearth.com. Just be sure to frame it as a "trip," so your children learn early that anything labeled as such can be an enjoyable outing, not something to fear. You can read more about building a vacation around a child's interest or passion in chapter 17.

AUTISM FRIENDLY VERSUS AUTISM CERTIFIED

Looking to start small? You'll find AutismTravel.com is a wonderful resource. Their website offers the most up-to-date listings on venues designated as Certified Autism Centers (CACs) or Advanced Certified Autism Centers (ACACs), and every family should download a copy of their latest *Autism Adventure Guide.*

The International Board of Credentialing and Continuing Education Standards (IBCCES) warns travelers of the distinction between autism-friendly businesses and those that are autism certified. While many travel destinations across the globe may claim to be "autism friendly," the IBCCES says that in reality they may be not properly trained and prepared to accommodate individuals on the spectrum because "they may not understand perhaps the most important part about understanding autism: that every person with autism is different. . . . This generally has an effect on social interaction and social perception, but the degree that a person is affected in each area (and even with different senses) varies significantly from person to person."[3]

To become autism certified, the IBCCES requires that a venue must be "committed to ongoing training and education to remain experts in the field. A Certified Autism Center (CAC) is a facility or organization in which at least eighty percent of their guest-facing staff is highly trained, fully equipped, and certified in the field of autism."[4]

Taking it to the next level, an ACAC necessitates, among other requirements, "forty hours of position-specific, customized training for staff and an annual onsite review of facilities by an expert that includes a comprehensive report with suggested changes or modifications to better serve individuals on the spectrum and other challenges, and sensory guides for activities."[5]

Many CACs also have sensory guides and undergo audits, but this may vary depending on size and type of location. Both the CACs and ACACs require renewal every two years, which means their staff must complete updated training and must continue to meet requirements in order to stay certified.

Many properties and travel vendors mentioned in this book are autism certified. To double-check, download the *Autism Adventure Guide*, which contains a complete listing.

Organizations other than IBCCES are now starting to offer autism- and sensory-inclusive certifications, among them the Champion Autism Network (https://championautismnetwork.com), KultureCity, which has just certified the Carnival Cruise Line as Sensory Inclusive™ Certified (https://www.kulturecity.org); Sensory City (https://sensorycity.org); and Autism Double-Checked (https://autismchecked.com). For example, the Lead with Love Training Company, developed and backed by the Champion Autism Network, offers seventeen different courses for online and onsite training, some given in Spanish, to certify corporations and organizations in one of three levels: Autism Supportive, Autism Participating, and Certified Autism Champion. At the time of this writing, they were in the process of certifying all forty-six of the Bluegreen Resorts properties around the United States (www.bluegreenvacations.com/resorts). You can read more at Leadwithlove.thinkific.com.

Before accepting the certified autism friendliness of any travel vendor, be sure to perform your own due diligence. What training does that certifying organization offer their partners? What criteria must a property or service meet to become certified?

Likewise, if a property or service in this book is designated as "AF" or Autism Friendly, be sure to research exactly what that means because the definition can vary widely. Perhaps check on social media for other special needs families' experiences with that vendor. You can never do too much advance preparation when it comes to your family's safety and well-being while on the road.

AUTISM-CERTIFIED ZOOS AND AQUARIUMS

As a short introduction to travel, consider a visit to one of these IBCCES-certified zoos and aquariums, especially if they are close to home. They are specifically trained to welcome both neurodiverse and neurotypical families.

- Arizona
  OdySea Aquarium
  9500 East Vía De Ventura, Suite A-100, Scottsdale, AZ 85256
  480-291-8000
  https://www.odyseaaquarium.com

- California
  Santa Barbara Zoo
  500 Ninos Dr, Santa Barbara, CA 93103
  805-962-5339
  https://www.sbzoo.org

- Georgia
  Georgia Aquarium
  225 Baker St NW, Atlanta, GA 30313
  404-581-4000
  https://www.georgiaaquarium.org

- Missouri
  Johnny Morris' Wonders of Wildlife National Museum and Aquarium
  500 W Sunshine St, Springfield, MO 65807
  888-222-6060
  https://wondersofwildlife.org

- Pennsylvania
  Elmwood Park Zoo
  1661 Harding Blvd, Norristown, PA 19401
  800-652-4143
  https://www.elmwoodparkzoo.org

- South Carolina
  Ripley's Aquarium of Myrtle Beach
  1110 Celebrity Cir, Myrtle Beach, SC 29577
  843-916-0888
  https://www.ripleyaquariums.com/myrtlebeach

- Tennessee
  Ripley's Aquarium of the Smokies
  88 River Rd, Gatlinburg, TN 37738
  865-430-8808
  https://www.ripleyaquariums.com/gatlinburg

- Texas
  Fort Worth Zoo
  1989 Colonial Pkwy, Fort Worth, TX 76110
  817-759-7555
  https://www.fortworthzoo.org

- Out of Country/Canada
  Ripley's Aquarium of Canada
  288 Bremner Blvd, Toronto, ON M5V 3L9, Canada
  647-351-3474
  https://www.ripleyaquariums.com/canada

CHILDREN'S PICTURE BOOKS ABOUT TRAVEL

The following books for young readers may pique an interest in travel. While suggested age groups are indicated, parents of neurodiverse children will have a better handle on the reading level of their child. You can find a wide variety of travel books for children of different ages, parsed into groups covering various destinations as well as different themes, at Book roo.com. Likewise, a Google search for "children's picture books about travel" will offer a bevy of suggestions, as will your trusted local librarian.

Little Traveler Board Book Set by Mudpuppy, illustrated by Erica Harrison (Mudpuppy Books)
> Suggested age group: 0–5 years
> Four mini board books that feature food, vehicles, landmarks, and animals from countries around the world.

Wherever You Go by Pat Zietlow Miller, illustrated by Eliza Wheeler (LB Kids)
> Suggested age group: 0–3 years
> A board book describing the escapades of an adventurous rabbit and his animal friends as they encounter the amazing world that lies just outside their homes.

Maisy Goes on a Plane: A Maisy First Experiences Book, written and illustrated by Lucy Cousins (Candlewick)
> Suggested age group: 2–5 years
> Young readers join a familiar character as she flies to visit her friend Ella and experiences every element of air travel.

Molly and the Magic Suitcase Series by Chris Oler, illustrated by Amy Houston Oler (COLOR Marketing & Design, Inc.)
> Suggested age group: 2–8 years
> This series tracks the adventures of Molly and her brother Michael as they travel to faraway places using a magic suitcase. The Olers created the series to "introduce children to cultures around the world" and not necessarily as travel guides.[6]

Family Trip (Peppa Pig) by Scholastic, illustrated by EOne (Scholastic, Inc.)
Suggested age group: 3–5 years
Peppa Pig travels with her family to Italy but keeps leaving her teddy bear behind everywhere she goes. The resolution may comfort readers with similar anxieties.

Richard Scarry's Busy, Busy World by Richard Scarry (Golden Books)
Suggested age group: 3–7 years
A short trip to each of thirty-three different locations and stories, including New York, Paris, Tokyo, Egypt, and Australia.

Let's Explore with Cor Cor by Cory Lee and Sandy Gilbreath, illustrated by Amanuel Moor (Cory Lee Woodard, publisher)
Suggested age group: 3–8 years
Children who identify with those who don't look like everyone else will appreciate Cor Cor's explorations of the world in a wheelchair, proving that everyone can and should travel.

National Geographic Little Kids First Big Book of the World by Elizabeth Carney (National Geographic Kids)
Suggested age group: 4–8 years
For children who enjoy facts, more than one hundred colorful photos and maps in this reference book are coupled with details about languages, landscapes, weather, animals, capital cities, mountains, deserts, and more.

Madlenka, written and illustrated by Peter Sis (Square Fish)
Suggested age group: 4–8 years
A good introduction to cultural exploration through staycations, Madlenka discovers the world by traveling around a city block filled with friends and vendors from different areas of the world.

A Ticket Around the World by Natalia Diaz and Melissa Owens, illustrated by Kim Smith (Owlkids)
Suggested age group: 5–8 years
A young boy takes a tour of thirteen countries across six continents that introduces readers to cultures from around the world.

A quick note: a picture book, such as those listed above, is not a Social Story. Read more about Social Stories and how they can assist with travel for both the neurotypical and the neurodiverse in chapter 6.

PARENT SPOTLIGHT—MICHELLE ZEIHR
OF BARRIE, ONTARIO, CANADA

Michelle Zeihr and her husband have two children. The eldest, Thessaly, who was nine at the time of this writing, is on the spectrum with ASD and ADHD. She is verbal and considered high functioning. The younger daughter, Eden, is five years younger than her sister and "seems to be neurotypical but I am having her monitored by an early intervention team, just in case," says Zeihr.

Though she started traveling with her children from an early age, Zeihr was concerned Thessaly would not be able to sleep without her exact bedtime routine: rocking in her rocking chair with the fan blowing, then sleeping in her crib. "I was also concerned about what she would eat. She is a very, very picky child. One of my [other] biggest worries was the plane ride and whether she would just cry the entire time and not understand she had to stay in her seat even if she wanted to run around."

It turned out Zeihr didn't have to worry, at least not about the flight. "I think people are afraid that an ADHD child would be rowdy and annoying, but people are surprised by how amazing Thessaly is on an airplane. I tell them she has done this since she was [a toddler], and I think that's why."

She says another misconception about spectrum travel is that all children need to stick to a consistent routine to successfully function—something she herself was initially concerned about regarding bedtime:

> We tried hard not to make her life so rigid that she couldn't adapt to a late bedtime or skipping her bath one night. We weren't the type where bath time was strictly at six, then dinner at seven, and bed at eight. Some nights, we'd go to the drive-in, and I would explain, "We are going to stay up late, skip our bath, and watch a movie. We can eat in the car."

Zeihr says this helped when it came to travel. "I would do the same with the trip, explain as much as I could . . . sometimes every step of the way."

To book a trip, she says she would "check all different websites for what we were after. All-inclusives were always great but sometimes, depending on the destination, we would use sites like HomeAway or Airbnb and rent a place."

Zeihr and her husband love to travel and wanted their children to experience different countries and cultures:

> I'm not one to lie on a beach or stay only on property at a resort. I like adventure so we try to pick a location that will give us easy access to local shops, restaurants, or car rentals to get around. We are animal lovers, and we help with many different rescues. When we go overseas, we try to find an animal rescue that we can visit and bring some things from home to donate. We play with the animals and take the dogs for walks. However, getting around can be difficult if the places we visit aren't stroller friendly.

Different cultures offer different cuisines, and that can cause problems, says Zeihr:

> We've [attempted] to show Thessaly what the different foods are like and although most times she's too picky to try them, she's always interested in trying a new ice cream or popsicle. . . . [Meanwhile], my neurotypical child, Eden, eats a variety of food and she's happy to try new things if they look good. We can almost always find food for her. For Thessaly we have to do a lot of extra planning and [find] backups in case there is nothing she can eat where we are staying.

Zeihr says that some of the travel modifications she made for Thessaly have benefited Eden as well. "Once Thessaly got older, we would find resorts with kids' camps. Thessaly would go for a couple of hours to have fun with other kids and that would give us a bit of one-on-one time with Eden."

Her main advice to other special needs parents: "Don't be afraid to go! The more you travel with them the easier it gets. They adapt so well over time. Now that Thessaly is older, she's the perfect travel companion and loves to fly. She looks forward to our trips. Also, having your child involved in something amazing, like snorkeling with you on a reef, is breathtaking! Remember, where there's a will, there's a way!"

CHAPTER 5 ACTION STEPS

1. Start small. Even a trip to a grocery store, a post office, or a garage sale can be turned into an adventure if framed correctly.

2. Capitalize on the diversity in your neighborhood to introduce cultural differences. Try out foods from different countries in local restaurants before planning a foreign vacation.
3. Structure local trips around your child's interests and passions.
4. Visits to local aquariums and zoos can be a great first step to touring; couple with "fun facts" if your child is so inclined.
5. An overnight at a friend's or relative's house can be a great precursor to a hotel stay.
6. Read picture books to your child that depict favorite characters and their travels.
7. When it's time to venture further, learn the difference between "autism-friendly" and "autism-certified" certifications and venues. Do your due diligence by learning what their certifications mean and what venues must do to earn them.

6

PRE-TRIP PREPARATION

"You should introduce the idea of a vacation early in the planning stages to give the child time to adjust to the idea of a trip away from familiar surroundings. Use pictures of the destinations to get them familiar with the location and hopefully [create] some excitement about doing something different but with some routine. Show your child each step of the vacation, packing, leaving for the airport, boarding the plane, arriving at the destination, meeting the staff, etc." – Jonathan Haraty, Adventure Luxury Travel, Whitman, Massachusetts

After the short staycation trial runs, you might consider taking a longer journey, further afield. All the parents and travel planners interviewed agree that preparation is the key to a successful holiday, whether for neurotypical or neurodiverse families.

"It is a fallacy that families with members [who have] special needs cannot take a cruise, go to a theme park, or [visit] an all-inclusive tropical resort," says Sandra Chu with The Enchanted Traveler in Rockford, Illinois. "This couldn't be further from the truth. It just requires more upfront research and a plan to minimize surprises."

To simplify things, we've divided preplanning into several components: research; planning as a group, including backup plans and creating safety protocols; packing; and previewing the journey.

RESEARCH EVERYTHING IN ADVANCE

Carl W. Johnson of Adventures in Golf (incorporated as Hamill Travel Services, Inc.) in Amherst, New Hampshire, calls advance research arguably "the key" to successful travel for everyone, be they neurotypical or on the spectrum. The dates of travel, the destination, how to access the location (direct flight or not), the length of the flight, how busy the airports will be (local or regional airports versus a larger Boston Logan or Los Angeles International)—these are all questions which should be asked and answered once the idea of taking a trip has been agreed to.

"Children get excited when families say they are going on a vacation," says Nicole Graham Wallace of Trendy Travels by Nicole—Dream Vacations in McDonough, Georgia. "Decide on a destination that is family friendly, one that has an amazing kids' program with trained professionals for neurotypical children as well as children on the spectrum," she says, emphasizing that all research should be done *before* verbally introducing the destination to the young ones.

Before booking any trip, Tara Woodbury of Escape into Travel in Leland, North Carolina, advises parents to familiarize themselves with what special needs programs and accommodations are offered both on flights and at their destination. "Most if not all companies [employ] someone whose job it is to help families with special needs," she says. Depending on the age of the child and where they are on the spectrum, she recommends considering something like an Autism on the Seas cruise (see chapter 11) or a resort with staff trained in helping kids on the spectrum stay safe and enjoy their vacation (see chapter 12). "Your typical cruise line or resort kids' club may not be able to accommodate your child, or at least not safely, which means you will be together twenty-four hours a day, seven days a week. That may not be your idea of a vacation," she says.

Jennifer Hardy of Cruise Planners in Kent, Washington, advises going a step further by seeking out and staying at destinations certified as autism friendly or Certified Autism Centers, as described in chapter 5.

Michelle Zeihr, a parent from Barrie, Ontario, Canada, always checks online reviews of where she'll be staying and the surrounding area:

> I also often look at the options on Google Maps and search for things that we can do off-resort with the kids. For example: see the ruins, visit an animal shelter, eat at a taco stand, visit the old market, or check out a museum. Food is always a concern: are they going to serve something aboard the flight? What food will the airline allow me to bring along? At the hotel: [Will there be] restaurants nearby that Thessaly will recognize? [Or] grocery stores where we can quickly grab some food? Other

concerns: Will the room have a balcony where my husband and I can sit while the kids sleep? Is there a bathtub or just a shower? Is there a kids' club that can accommodate Thessaly?

One of the benefits of advance research is the ability to weed out upsetting stimuli, something that should always be a consideration when choosing attractions to visit, according to Taylor Covington, who writes for *The Zebra*:

> [Children can be] overly stimulated by crowds at times [so] a trip to an amusement park, where strict schedules are a necessity, may not be the best choice. Choose a trip to a relaxing destination with unhurried schedules, like the mountains or the beach [or] choose a destination based on your child. . . . Some children may find the heat of the sun and the scratching of the sand to be too troublesome for a beach trip. Some children may love amusement parks, even with their neurodiversity.[1]

PLAN AS A FAMILY

Planning a vacation should be a group affair, explains William Johnson of Details by Bill LLC in Germantown, Wisconsin. "Discuss the trip with the entire family walking through each step, describing the expected environment. Visit Autismspeaks.org for tips and information," he counsels.

Belinda S. Stull of Travel Leaders/All About Travel, LLC, Hagerstown, Maryland, agrees:

> I think it's great to have children involved in the preparation of their vacation. In some cases, it's as simple as talking to them about the destination or counting down the days to their trip. In other cases, I have had children very involved in helping their parents plan the day-by-day activities or suggesting certain activities or places they would like to see. Helping to pack their suitcases is another way some children like to get involved.

When including children in the process of selecting a destination, Jennifer Hardy suggests offering them two or three options, and advises parents to

> keep your child's interests and hobbies in mind when planning. If your child really loves trains, be sure to include train travel or a railway

museum on your itinerary to keep your child engaged. Create a schedule but build in extra time for flexibility. Review the schedule with your child at least three times before your trip begins so that they know exactly what to expect and in what order. Remember to stick to this schedule as best as you can. This will help all children to remain calm and alleviate any anxiety.

Sarah Salter of Travelmation dba Sarah Travels Happy in Fayetteville, Georgia, believes that anytime you get input from children, a trip goes smoother:

As a travel agent I offer to hold a Zoom [meeting] with the kids to find out what's on their list so they feel involved and important. It gives them a stake in the success of the trip. [You should] always use the rule of two right answers when planning [a selection of two choices, both acceptable to parents.] That makes a child feel empowered and included, and a parent isn't afraid of a "wrong" answer.

Another way to involve the child in the decision-making process is to "let them help develop a storybook outlining all the different aspects of the trip: when you will be leaving, how long you will be gone, where you will be stopping; and the schedule for each day," according to Sally J. Sutter of Mermaid Travel in Canal Winchester, Ohio.

Angela Romans of Rejuvia Travel in South Bend, Indiana, advocates leaving time for sensory timeouts and sensory friendly activities when planning the itinerary. She also advises packing a go-to bag (a bag or backpack containing emergency items such as acceptable food and drink, change of clothes, sensory items, familiar toys, etc.),[2] having an exit strategy in place, and setting up a signal for the child to use when things become too overwhelming. This includes making sure the hotel, restaurant, or venue staff are aware of the signal too, so they can help if needed.

All these tips are designed to alleviate stress, a vital precaution because, according to Taylor Covington, "children with autism can pick up on the stress of others . . . and the more stressed you are, the more your child will struggle. Whether you need to plan extra time, need down time for yourself, or need to bring an extra caretaker . . . to help with your child, do what you can to reduce your own stress while planning and executing your trip," she warns.[3]

HAVE A BACKUP PLAN

Experience has taught Nicole Thibault of Magic Storybook Travels in Fairport, New York, the importance of formulating alternative plans, especially when her son faces situations requiring an action outside his comfort zone:

> A few years ago, we planned an activity that had our family crossing a suspension bridge. I spoke with the tour guide ahead of time and explained that considering my son's autism and anxiety issues, I wasn't sure if, when the moment came, he would be able to cross that bridge. The guide and I came up with a backup plan, that if he froze and couldn't cross, there would be a golf cart waiting for him, ready to take him to the end point, and we could continue from there. Ultimately, he surprised us all and crossed with no problem, but the backup plan was in place, just in case.[4]

CREATE SAFETY PROTOCOLS

Unless otherwise indicated, all tips in this section are borrowed with permission from Taylor Covington's article in *The Zebra*, titled "Traveling with Autism: How to Handle Safety, Transitions, and Time in Transit."[5]

- Make sure you have a plan if your child has wandering tendencies. Using tether backpacks can be helpful, as it gives your child a measure of freedom while ensuring you know they are always close by.
- Consider a GPS tracker. Products like AngelSense (a GPS tracker designed for those with special needs) give parents the peace of mind of knowing they can find their child if they wander off.
- Use a medical alert bracelet if your child has speech concerns or other considerations that emergency personnel should know about if an accident occurs. On the bracelet, include an emergency number for the phone you will have on you during your travels.
- Have an ID kit with you that can be used to help find your child in an emergency. A recent photograph and an accurate description, including current height and weight, can help emergency responders locate your child should you become separated. The National Child Identification Program offers kits you can use to

easily create an ID packet for your child, including a fingerprint and DNA sample, which is a good starting point.

- Teach your child what to do if separated. From staying within a safe area to identifying safe strangers, like police officers or store workers, give your child skills to handle dangerous situations in case you are separated. Of course, this needs to suit your child's abilities and understanding, but it is something to practice at home.

- "I always made sure that when the kids were younger, they had identification in their pockets with the name of the hotel, our phone numbers, and money for a cab. When my son was six, he jumped on the train in Paris and the doors closed. He had no language skills and was severely dyslexic, so that being said, preventive measures are wise with young children," says Dawn Baeszler of AAA Travel, Lawrenceville, New Jersey.

- "Consider temporary safety tattoos. These temporary tattoos can be placed on your child's arm [featuring] your name and phone number, increasing the chance that your child will be safely returned if lost and separated," suggests Nicole Thibault of Magic Storybook Vacations in Fairport, New York.[6]

PACKING FOR PEACE

What you pack can make a huge difference when preparing to travel with children on the spectrum. Avoiding irritating clothing should be at the top of the list. Consider the experience of Delight Dolcine, a parent from Brooklyn, New York:

I remember when Xenia went to Florida on our first vacation. She was three years old. She looked so cute in her Dora shorts and bathing suit. I love thong-style sandals and of course I had to put my "mini me" in my favorite-style shoes. But every time we placed her down to walk, she would cry, and then, when I picked her up, she would be her normal playful self. I'd put her down and she'd cry again. As soon as we got to the room, she flung those sandals off and rubbed in-between her toes. That is the first time I realized that she was sensitive there. I felt so bad forcing her to keep her shoes on, thinking it was a normal toddler power struggle. I began researching the best summer shoes for children on the spectrum. I wanted something flexible, waterproof, and cute. So now I buy her supportive hiking sandals and if she gets wet, they dry easily. I like the brands Sorel and Merrell for her.

"I tend to overprepare when we travel," explains Michael Sokol, a parent from Longmeadow, Massachusetts. "Nate likes his devices—iPad, iPhone, Switch DS, etc. They entertain him and distract from any chaos around us. I will bring a couple of devices, extra charging powerpacks and cords, snacks, drinks, and noise-cancelling headphones. We also travel with a laptop," he says.

With all children, familiarity equals comfort. That's why Brenda Revere of C2C Escapes Travel in Bedford, Indiana, recommends packing "that favorite T-shirt and comfortable clothing and shoes. Make sure you have that special blanket or stuffed animal with you to avoid meltdowns."

Other must-haves for parents with children on the spectrum include the electronics and toys the child uses at home, according to Sally J. Sutter, who lists "noise-cancelling headphones with music and games, fidget toys, favorite toys and activities, weighted blanket, water, and snacks" for a parent's survival bag. On her list, Tara Woodbury recommends toys for children who enjoy tactile stimulation, like Play-Doh or a squishy. As for comfort items, she suggests not washing them right before your trip. "The familiar smell of home may help your child feel more at ease," she says.

"Ask your children if they would like a small, kid-sized backpack that they can wear and carry," advises Natasha Daniels of Anxioustoddlers.com. "Tell them they can put a few objects in their bag. Make sure to keep it light so they will be able to carry it themselves. Also, this is not the time to bring a bunch of new clothes that have never been tried and tested by your picky or sensitive children. If the tags are still on them— leave them at home!"[7]

What if the kids want or need to bring more toys than the family's luggage will permit? Dr. Christopher Scott Wyatt, a Texas-based college professor on the spectrum and the creator and host of the Autistic Me Blog and Podcast, has come up with a great solution—UPS. Three days prior to any trip he and his wife take with their two neurodiverse daughters, he ships their favorite toys, like Koda, the polar bear, and Kobu, the red panda, to the family's intended destination. How do the girls deal with missing these animals while they are enroute? "They choose which animals go and help with the packing. They understand that the toys are taking a mini vacation on their own, but will be reunited with them soon," Wyatt explains.

There's more to packing than just including the right clothing, toys, and electronics, though. "The noises in a hotel can be uncomfortable for a child with autism," says Covington. For that reason, she suggests that parents

> pack a white noise machine or fan that you can use in the hotel to mask the noise as much as possible . . . also, foods. Many children with autism are picky eaters. . . . Keep hunger at bay, especially when unusual food options present themselves, by packing some foods that your child considers "safe." Keep in mind that airports have regulations about which foods and liquids are allowed in carry-on baggage, so read up on those before you head out. . . . And toiletries. While most hotels provide soap and shampoo, your child may do better with the smell and texture of the items you use at home regularly, so bring your own.[8]

PREVIEW THE ADVENTURE

You can't just spring a vacation on a neurodiverse child. So says Sarah Marshall of TravelAble in Midlothian, Virginia, who explains:

> For neurotypical children, you may be able to surprise them [with a trip] the night before and they will be excited. But for children on the spectrum, this could be the source of tremendous anxiety. So, I prepare a personalized story for the child, I send books or resources about getting on the airplane, what the hotel will be like, and so forth.

"To give my daughter Xenia a hint that we'll be traveling, I let her pack with me a few days in advance and roll her suitcase around the house as if we were in the airport. That's how I know she is aware a trip is coming up," says Delight Dolcine.

"We need to tell our daughter who's on the spectrum what to expect a day or so before the trip and [also] a day or two before we go home, unlike the younger one," says Kristen Chambliss, a parent from League City, Texas.

Betty J. Burger, recently retired from BGB Travel and Leisure, LLC, in St. Marys, Georgia, suggests role-playing parts of the trip at home to lessen anxiety, while Jennifer Hardy recommends locating or creating Social Stories (described later in this chapter) "to help [the] child understand the changes to their routine and [manage] expectations. Be sure to review these with your child at least three times before you depart and bring them with you for your child to review as you check each portion of your trip off the Social Story progression."

Because many children with autism spectrum disorder (ASD) have sensory issues, if planning a beach vacation, it may be helpful to introduce what sand feels like, advises Lisa Bertuccio of Vacation Your World in Otisville, New York. "You can buy clean sand at any home improvement store and have the child touch it and put their feet in it. We take for granted that neurotypical children will just adapt to that new feeling of sand in-between their toes, but it may be too much for a child with ASD."

Wings for Autism® is a great tool for families on the spectrum to familiarize themselves with the airport travel and navigation process, says Tara Woodbury. "Children have a chance to practice arriving at the airport, obtaining their boarding passes, checking bags, being screened at the TSA security checkpoint, and boarding the aircraft. It's like a dress rehearsal for their flight. Similar programs are available throughout the country." You can read more about programs like Wings for Autism in chapter 8.

SEEK THE ADVICE OF SPECIALISTS

"Talk to your child's doctors and therapists about your plans," advises Woodbury. "They may have some great ideas to help make travel go more smoothly. If your child needs anxiety medication to manage day to day, you may ask about getting an as-needed dose for the plane ride."

"If you work with therapists, especially occupational therapists or Applied Behavior Analysis therapists, have them help you prepare resources and work one-on-one with your child to understand the upcoming trip," suggests Sarah Salter. "A nonverbal child would need a different approach than a child who can read and communicate at a high level. Incorporate the lead-up to a trip into your existing behavioral strategies such as checklists or picture prompts. Put together a 'what to expect' booklet with maps, scenarios, and pictures," she says.

These preparations are a good first step. In upcoming chapters, I'll focus in on deciding where to travel, how to get there, where to stay, and what to do once you've arrived.

SOCIAL STORIES™

Social Stories are a method autism consultant Carol Gray developed as an engaging and interactive way to help those on the spectrum cope with new or problematic situations. According to Beth Arky in her article for the Child Mind Institute called "Tips for Traveling with Challenging Children," the stories, which are

> written from the child's point of view—use narration, as well as photos and drawings, to guide the child through an experience, preparing him for social interactions that might be required and prompting desired responses. A story for a child on the autism spectrum about a trip to Florida might include travel details, things he might be nervous about, people he might meet, reassurance that his parents will be with him all the time, and activities he likes and can anticipate enjoying.[9]

According to Autism Inspiration, the story should be written from the perspective of the child and include pictures as well as at least four types of sentences:

- Descriptive: including opinion-free facts about where the situation is taking place, what [the people in the pictures] are doing, and why they may be doing it
- Perspective: describing the feelings, thoughts, and perspectives of others
- Directive: giving a suggested example of how to handle the situation or problem
- Control: usually composed by the [child] to identify a personal coping strategy for the situation[10]

The internet is filled with guides and templates of how to write Social Stories, and many proponents see them as helpful not only for children with autism, but for those with many other developmental challenges as well as for neurotypical children. "To me, curling up with my kids and reading a cute little book is always fun, and if there are lessons involved, even better. Social stories really are a win/win," says Alethea Mshar in her article, "Special Needs Parenting: What Is a Social Story and How Can I Get One?"[11]

You can find several templates and examples on the Autism Speaks website, including one about flying, created in conjunction with Jet-

Blue Airways at https://www.autismspeaks.org/sites/default/files/Jet%20 Blue%20teaching%20story%20final%209-23-19.pdf.

You can also download an app to make your own Social Stories. The Social Story Creator & Library app allows you to write and illustrate your own stories for different situations. As it can reside on your cell phone, it's something handy for teaching on the go. Some versions are free; others have a charge. Currently available only for iPad and iPhone, you can download it from the Apple App Store or at https://apps.apple.com/us /app/social-story-creator-library/id588180598.

PARENT SPOTLIGHT—DELIGHT DOLCINE

Delight Dolcine of Brooklyn, New York, refuses to let anything stop her from exploring the world. The parent of a child with ASD and moderate intellectual disabilities, she's been traveling with her daughter since the child was three, despite worries over tantrums and safety—what would happen if nonverbal Xenia got lost? What about predators and child abduction? Yet she wanted to relax and began to define vacations as "self-care." She also says she loves her travel photos and memories and so she ignores the advice of naysayers who insist that "'special needs children won't know what's going on [while traveling]; therefore, it's a waste of money to take them.' Some people have even said that I should use that money [for] therapy/services instead."

What gets Dolcine and her daughter through trips is "both of us being well rested and fed. Previewing where we'll be staying, either through videos or pictures. Packing for the trip. If you have a partner or relative who can come, please bring them [because] it is much easier, and you get better pictures."

Traveling with Xenia is an ongoing learning process, she admits:

> On our trip to Buffalo last November, Xenia kept leaving her suitcase everywhere in the airport. This was our first trip with just the two of us flying together, so I was focused on not losing her, not missing our flight, and making sure she was comfortable. (She has limited vocabulary and [can't] easily tell me when something is bothering her.) So, we're running through the airport, and I would turn around and her suitcase would be by the customer service desk, and I'd have to drag it back. Or she would leave it in a bathroom. For the most part I was watchful but

there were at least two times when passengers were like "Miss? Miss? Your suitcase!" She was unbothered, feeling no connection to worldly things like the clothes and shoes in the suitcase. But if we lost her Play-Doh, she'd have a conniption! I am still trying to figure out how to keep her focused on her suitcase in the airport for next time.

CHAPTER 6 ACTION STEPS

1. Research everything about the trip in advance of introducing the destination or venue to a child on the spectrum.
2. Make sure your destination has special needs programs in place and, preferably, has an autism-friendly or autism-certified designation. This was discussed in chapter 5.
3. Once the location is chosen and vetted, invite children to participate in the planning, offering them a selection of choices that you've already preapproved. Developing a storybook together might help, as well as deciding on a signal strategy so kids can let you (and travel staff) know when anxiety hits.
4. Prepare the child for the adventure by using Social Stories, schedules, and role play.
5. Introduce unfamiliar items at home that may be encountered when traveling, like sand. Your child's therapist and doctors may also be able to help by providing both advice and as-needed medications.
6. Make sure the experience is as stress-free as possible, both for you and your children. This means avoiding whatever external elements may irritate the child, such as crowds.
7. Always have a backup plan in place.
8. Safety first. Consider whatever measures might be necessary to keep children with you, and to locate them again in case of possible separation, such as tethers, GPS trackers, medical bracelets, ID kits, keeping identification with the child, and teaching the child what to do if separated from you.
9. Pack for comfort and bring whatever is necessary to keep the child occupied and well-fed. Bring along the soaps and shampoos they're used to, as well as clothes that still harbor the familiar scent of home.

7

FOREIGN VERSUS DOMESTIC?

"We started off traveling domestically only due to normal concerns of catastrophic events. If you're domestic, it's just easier because of access to medical care and most people speak the same language, and the currency is [familiar]. We traveled to Bermuda in 2019 but that was still very safe because they speak English and have the same comforts as New York. I did want to travel to Luxor in Egypt for my birthday, but because of the pandemic, we went to Universal Studios instead.

"I decide on the destination, not because it's domestic or international, but mainly based on safety and crime rates. I also try to travel during the off-season, when there are fewer tourists and less noise, plus in general, it's cheaper. I check if the hotel is an accredited ASD facility like Sesame Place in Pennsylvania and make sure they offer reasonable accommodations or modifications. Reading reviews and looking through hundreds of photos and YouTube videos also plays into my decision-making process."
– Delight Dolcine, parent, Brooklyn, New York

Many of the families on the spectrum interviewed said one of their primary goals when traveling abroad was to introduce their children to different cultures. But international travel creates additional trials: adapting to new foods, languages, currencies, and potential health concerns that may require vaccinations.

ANYTHING'S DOABLE

For children who are already sensitive to changes in their environment, do the benefits outweigh the challenges? Dr. Tony Attwood, an autism spectrum disorder (ASD) expert, believes that they do. "In general, I have found children with ASD have adapted to vacations, especially abroad, remarkably well," he says. "As these are great changes in that person's life, they may be more easily accepted than minor changes such as rearranging the furniture in a bedroom. Upon their return home, for the family to return to their usual routines, the person with ASD will need to be congratulated as quickly as possible for coping so well with the situation."

The consensus was that anything's possible with the right preparation, though it depends on the travelers involved. "I do not feel it is necessary to limit travel based on destinations as long as you invest the proper amount of time in prep work and seek help from a qualified travel planner," says Sarah Salter of Travelmation dba Sarah Travels Happy in Fayetteville, Georgia. "Like most things in life, accommodations will need to be made, but it is essential to start from a perspective of how your child 'can' access an experience or destination [instead of automatically assuming] that they 'can't.'"

The choice between traveling internationally versus staying within the United States can be a tough call, according to Jonathan Haraty of Adventure Luxury Travel in Whitman, Massachusetts:

> Do you break them in with a short trip to a local destination or do you look at flying or sailing internationally? If you live near a theme park that accommodates children on the spectrum, this type of domestic trip may be an option for a trial run. Also, cruising is a good option; [many involve] short flights to the departure city and local ports. Cruise lines like Royal Caribbean and Carnival have trained staff who are certified to work with children on the spectrum. This will give parents some well-deserved alone time while their child is with the staff. (Read more about theme parks in chapter 14 and cruising in chapter 8.)

"It all depends on [the children's] age and specific developmental and behavioral challenges as well as their past experiences with travel," says Chandra Nims Brown of Cruise Planners—Global Travel Pros in Williamsburg, Virginia. "Both can successfully be done under the right circumstances." To that end, Sarah Marshall of TravelAble in Midlothian, Virginia, recommends a step approach. "If your child has never been on an airplane, a twenty-hour flight to Asia should NOT be their first. Ease them into it with smaller trips, then a two-to-three-hour flight to a domestic destination, then longer," she suggests.

STAYING DOMESTIC

Domestic is the way to go, "especially if the parents are not used to leaving the country," says Starr Wlodarski of Cruise Planners—Unique Family Adventures in Perrysburg, Ohio. Some of the reasons? Betty J. Burger, recently retired from BGB Travel and Leisure, LLC in St. Marys, Georgia, warns of the extended length of international flights during which a child would need to remain confined in an unfamiliar space. "A long flight overseas is not the time to test how they will do with flying," agrees Tara Woodbury of Escape into Travel in Leland, North Carolina.

Burger's other considerations include language barriers and variable availability of medical treatment. Crossing into different time zones is also an issue, according to Holly Roberts of Frosch Travel—The Travel Center in Hanover, Massachusetts.

Of course, cross-country travel within the United States can also involve lengthy flights and time changes, and sometimes, an international flight might be shorter than a domestic one, as Carl W. Johnson of Adventures in Golf (incorporated as Hamill Travel Services, Inc.) in Amherst, New Hampshire, points out. "Our first family trip was to Bermuda which did require extra steps in the process (e.g., going through customs). But we felt the additional steps upon arrival into an international destination were going to be okay since the flight itself was so short (under two hours)," he recounts.

"For your first trip, it is generally a better idea to stick closer to home especially if your child is on the more severe end of the autism spectrum," says Tara Woodbury. "While you can travel abroad with someone with ASD, many countries don't offer the same protections and regulations as the United States. When you are ready to travel internationally, check if the country or province has an autism advocacy board. They can be a great contact for you. Always do your research before heading overseas, no matter what your family makeup," she advises.

Another proponent for starting domestically is Sally J. Sutter of Mermaid Travel in Canal Winchester, Ohio. "As much as I think it is important for children (and adults) to experience different countries, in extreme cases, the overload of stimulation might become too much for a child, and a trip would need to be shortened. This is more easily done with domestic travel," she says.

DON'T RULE OUT INTERNATIONAL TRAVEL

Touting the benefits of international travel, Nicole Graham Wallace of Trendy Travels by Nicole—Dream Vacations in McDonough, Georgia, says she recommends venturing abroad because "it gives [families] a chance to meet different types of people from around the world and make new friends."

Other Certified Autism Travel Professionals espouse the advantages while offering caveats. For example, Lisa Bertuccio of Vacation Your World in Otisville, New York, suggests that language barriers could be an issue internationally, "especially if an unexpected behavioral issue arises or if special accommodations are needed during an excursion [when] you would want to communicate those needs effectively." One solution could be hiring a bilingual tour guide.

"Depending on the destination, the lack of familiarity can be overwhelming. Not all destinations have programs in place for families on the spectrum and it can be difficult to navigate between autism friendly and autism certified. It would be a good idea to help familiarize your child with your vacation destination before arrival. Videos are a great and fun resource," says Angela Romans of Rejuvia Travel in South Bend, Indiana.

Domestic and international vacations are equally attainable, but you will need to consider the additional steps involved with an international vacation as well as the lack of understanding of autism in some countries, cautions Jennifer Hardy of Cruise Planners in Kent, Washington:

> It is important to work with a travel advisor who can recommend destinations and suppliers that will be accommodating and understanding of your family's needs. Personally, I have taken my four children with invisible disabilities to international locations, mostly via cruise ship as it provides some familiarity of home and allows us to maintain a decent and predictable routine.

"We like both foreign and domestic travel but probably a bit more planning goes into international travel," admits Michael Sokol of Longmeadow, Massachusetts, the parent of a child with ASD. "For us, it's more like fly or not fly? I also like to have a go-to contact in the country we are visiting. This could be a friend, a friend of a friend, or someone my travel agent may know."

"Research and planning are key here for all families, but especially for those with sensory issues who may be put off by the different taste of food, the different sound of language," advises Sandra Chu of The Enchanted Traveler in Rockford, Illinois. "Trying to simulate some of these experiences before travel is a good idea. Breaking up a trip might be recommended if there is a big time change—maybe staying overnight halfway to adjust, for example."

It's clear that with the right preparation, there's a world of possibility out there for travelers on the spectrum. For those parents interested in traveling abroad, many of the books listed in chapter 5 will introduce children to the concept. For additional suggestions, google books that address various destinations and modes of travel and/or ask for referrals from librarians.

AUTISM-FRIENDLY/CERTIFIED
INTERNATIONAL RESORTS AND TOURS

Patronizing autism-certified or autism-friendly hotels, resorts, and tour companies can help prevent problems before they start, or at least minimize their impact. Unless indicated otherwise by the code AF for autism-friendly, the following properties are Certified Autism Centers or in the case of Beaches, Advanced Certified Autism Centers, which have earned their designation from International Board of Credentialing and Continuing Education Standards. (Read more about the certification process in chapter 12.)

Resorts

- **Belize**
  Ramon's Village Resort
  Coconut Drive, San Pedro, Brazil
  U.S. Office: 800-624-4215
  https://ramons.com/
  Styled after the Tahitian cottages on the Polynesian island of Bora Bora, Ramon's Village Resort describes itself as a sanctuary nestled in a tropical garden of royal palms, bougainvillea, lilies, hibiscus, and other tropical flora. Activities include scuba, fishing, and lounging by the pool or on the beach. Local touring can also be arranged.

As per their entry in AutismTravel.com's *Adventure Guide 2021*: "For years, Ramon's Village Resort has had guests both new and returning who travel to the resort with children of special needs. When we learned many children with autism are calmed by water and that the [International Board of Credentialing and Continuing Education Standards] and PADI had developed a program to help dive instructors work with these individuals, we felt compelled to be a part of such an inspiring program." – Steve Sherwin, Vice President of Operations.[1]

- **Beaches**

Beaches Resorts, Ocho Rios
Old Main Road, Ocho Rios, Jamaica
888-232-2437
https://www.beaches.com/resorts/ocho-rios

Beaches Resorts, Negril
Long Bay, Negril, Jamaica, Norman Manley Blvd, Jamaica
888-232-2437
https://www.beaches.com/resorts/negril

Beaches Resorts, Turks & Caicos
Lower Bight Rd, The Bight Settlement TKCA 1ZZ, Turks & Caicos Islands
888-232-2437
https://www.beaches.com/resorts/turks-caicos

"If you are willing to take the plunge on flying to an international destination or have an older child on the spectrum or a young adult, I would suggest staying at a Beaches Resort. Beaches was the first land-based destination resort certified as autism-friendly and all their staff has been [specially] trained," says Jonathan Haraty of Adventure Luxury Travel in Whitman, Massachusetts. "They even have one-on-one services for your child. Their resorts are large with plenty of activities to accommodate all children and teens and even adults on the spectrum." (Read more about Beaches in chapter 12.)

- **Peru**

Juve Travel Peru
A.P.V. Virgen Concepcion W-2-B, Santiago, Cusco, Peru
51-986016545
https://juvetravelperu.com/contact/

Juve Travel Peru says they are extremely passionate about travel and specialize in making challenging destinations accessible to everyone who is able. According to their website: "Whatever you want to experience in Peru, we can help you create an itinerary that takes you one step beyond your everyday, one step beyond your normal limits, and one step closer to your dreams."[2]

- **Portugal**
  4 All Senses
  Rua da Prata, 80, 1100-415 Lisboa Portugal
  351-963-364-359
  https://portugal4allsenses.pt/
  4 All Senses specializes in Portugal tourism for people with special needs. Families can choose between having the company provide transportation or need-based accommodation, or they can join one of the company's thematic tours. The company can also arrange private tours of Lisbon and other popular destinations that will accommodate any needs a child with autism might have.

Cruising the World

Cruising offers the best of both worlds: a short flight to a departure port, and minimal upheaval once there—no need to pack and unpack with each port of call. Royal Caribbean, Celebrity, Norwegian Cruise Line, Disney Cruise Line, and Carnival Cruises are among the cruise lines offering autism-friendly cruises to the Caribbean and beyond. (Read more about cruising in chapter 11.)

PARENT SPOTLIGHT—MICHAEL SOKOL, LONGMEADOW, MASSACHUSETTS

Michael Sokol and his wife are the parents of Nate, who was born prematurely at twenty-five weeks and later diagnosed with ASD. They started traveling with Nate to Cape Cod when he was around eighteen months old:

> We travel for vacation, but we also want to expose Nate to as many experiences as possible. Usually, our main challenge is just getting to our destination. Keeping Nate occupied on a four-hour flight is a mission, as well as making sure transportation is set and ready to go upon arrival. Less stress on him is less stress on us. I try to prearrange as much

as possible so Nate "knows the plan" and this helps with transitions. Cruises are great because you can preplan all of your shows, activities, and meals. Royal Caribbean has a fantastic Autism at Sea program.

The biggest misperception is that you can't travel with your special needs child, [or worry that it will be too] complicated and stressful: "What will people think if my child with autism acts out while on vacation? My special needs child will not enjoy the trip. It's too difficult to travel without all our supports, no one will help us."

I like to plan trips that give us a variety of activities. Usually the more experiences the better, but I also include downtime. My son is big on "first this . . . then that," so the smoother the transition, the better.

Sokol says he subscribes to a "don't care what other people think" philosophy:

Don't be embarrassed because your kid is "stimming" [stimming usually refers to specific behaviors by a person with autism that include hand-flapping, rocking, spinning, or repetition of words and phrases], or your special needs child is having a meltdown. There are plenty of "typical" children and some adults I have witnessed having meltdowns ten times worse than my son with autism.

I'm lucky because Nate travels fairly well which I attribute to [us] dragging him all over starting [from] a young age. But there were times I've needed assistance or some special treatment to sit up front at a show, get in before the crowd arrived, or even find a quick way out after a concert—all you need to do is ask. The worse thing that can happen is that you are denied, but I have found most if not all people will help.

The last concert we took Nate to was Maroon 5. He loves music but not the crowds. Nate wore his headphones to help reduce the loudness. Towards the end of the concert, I asked an usher if there was a quick way out. She took notice of Nate's headphones and I think some of his stimming. Smiling at Nate, she said, "Follow me." The next thing we knew, we found ourselves heading to an elevator behind the stage. The last song had ended, and the band was also headed toward that elevator. So, my point is, ask for assistance because the outcome may be better than you expected.

CHAPTER 7 ACTION STEPS

1. Consider the child before deciding how far to fly, or if you should attempt to visit a different country. Are they able to sit for long periods of time? Have trouble sleeping when changing time zones? Adapt well to changes in language and food?

2. Start small and then build up to longer trips or farther destinations, all with a positive attitude.

3. Remember that cruising might be a way to experience an international destination without flying or with a shorter plane ride, depending on your starting point.

4. Break up long journeys halfway through if necessary to acclimate to the time change.

5. Before venturing abroad, consider the extra steps involved, such as going through customs.

6. If traveling abroad, look into hiring a translator or bilingual tour guide.

7. Check to see if the destination country or province has an autism advocacy board and contact them in advance with questions or when seeking suggestions. Also ask your friends, relatives, or travel advisor for a local contact wherever you are headed.

8. Use videos to preview the destination. Also try to sample the foods and the sound of the language before leaving home.

9. Seek out autism-friendly and autism-certified hotels and venues at your destination.

10. Work with a travel advisor who specializes in autism travel when making plans. They can handle a lot of these steps for you.

11. Upon your return home, congratulate your child for how well they adapted to the changes in their routine.

Part III

MODES OF TRAVEL

8

NAVIGATING AIRLINE TRAVEL

"My son was diagnosed with pervasive developmental disorder not otherwise specified (now referred to as 'atypical autism') at age five and was an extremely picky eater until he was about thirteen. . . . He would lose weight during every trip and often got sick as well. Once, when we got home, he got on his hands and knees in his room and kissed his floor, he was that happy to be home.

"I had to fly with him to Vegas when he was five, and I was doing it alone. To make sure we survived the flight, I packed six Tupperware containers with his food and his snacks, a backpack of drawing materials, and a little portable DVD player. Now you can bring an iPad, but [back then, they] didn't exist. My goal was to keep him occupied with either eating, drawing, or watching movies for the entire six hours. It worked! So that's my advice: bring as much familiar food as the carry-on rules allow, and remember, the iPad is your friend."
– Anonymous mom, Rockland County, New York

Many families are intimidated by the idea of bringing an anxious, inflexible, or neurodiverse child aboard an airplane. And it's clear why: hours stuck in a confined space with popping ears, unfamiliar sounds, smells, and strangers, with no possibility of an early exit. Boredom guaranteed to set in after only a brief time. Angry looks from bystanders, should a meltdown occur. As Marian Speid of Dream Vacations in Fair Lawn, New Jersey, warns her clients, "Being on a plane is often overwhelming for anyone with or without special needs."

"Many of the tips for air travel are very effective even for children who are not on the spectrum," agrees Lisa Bertuccio of Vacation Your World in Otisville, New York. "These include booking direct flights, packing essential items to keep the child from getting bored, choosing aisle or bulkhead seating to give the child more legroom in the hopes of minimizing seat kicking, and taking their shoes off once seated."

Of course, for some whose sons or daughters are obsessed with aviation and may even be plane spotters, getting a child on an airplane might not be that difficult.[1] It may, in fact, be the highlight of their trip. But for others, it might not be as easy. By following the strategies of those families who have experienced success, flying can open your world to exploring novel destinations and enjoying new adventures together. Here are their methods, broken down into several small, easily digestible steps.

INTRODUCING THE IDEA OF FLIGHT AND PREVIEWING THE EXPERIENCE

As with many new concepts, parents and Certified Autism Travel Professionals (CATPs) recommend introducing the idea of airplanes, airports, and flying long before travel commences. Kristen Chambliss, a League City, Texas, mother of a child on the spectrum explains, "We read [our daughter] Social Stories so she knows what to expect and have gotten her books about the airport." (More about Social Stories can be found in chapter 6.)

Jodi Heckman-Caliciotti, a special education teacher and behavior specialist from Rockland County, New York, is a strong believer of helping children on the spectrum preview experiences in a non-threatening environment before they happen in real time:

> I remember back when I worked in classrooms teaching social skills. I knew one of my students was going on an airplane for the first time. I showed pictures of real planes and described the roles of pilots, flight attendants, passengers, and baggage handlers. Then we set up the room like a plane and took turns performing different roles. We played airplane sounds and spoke on microphones too. All the kids loved it so much and my student did awesome on his first plane ride!

"For children on the spectrum, it is very important to practice [beforehand] what air travel will be like such as removing shoes during security," says Lisa Bertuccio. To give youngsters on the spectrum this type of practice, the Charles River Center (an affiliated chapter of The Arc), in collaboration with the Massachusetts Port Authority, developed a program called Wings for Autism. The goal was to help children with autism and other disabilities practice how to navigate the airport, including checking in, getting through Transportation Security Administration (TSA) security inspection, and boarding an aircraft. The inaugural program was at Boston Logan International Airport and is still held twice a year. Wings for Autism is now currently modeled at sixty-nine U.S. airports. Massport also teams up with individual airlines and TSA to offer personal meet-and-assist services for those families who need extra help, says Bernice Freedman, their spokesperson.

"Even if you can't book a Wings program, airlines such as JetBlue will allow you to make an appointment to visit the plane and talk to pilots while you are at the airport for the tour," says Jonathan Haraty of Adventure Luxury Travel in Whitman, Massachusetts. "Please take advantage of [their Blue Horizons for Autism program] so you can see where you may need to make accommodations for triggers that occur."

Michael Sokol, the parent of a child on the spectrum from Longmeadow, Massachusetts, witnessed the airline's program firsthand. "We first visited JetBlue a week before our trip; walked Nate through the entire process and onto an aircraft. The day of travel, we were assigned an escort who checked us in, brought us through the security search, and onto the plane. She stayed with us until the flight departed. I cannot say enough about how great JetBlue treated us."

Similar programs include Autism Takes Flight at Wilmington International Airport in Delaware and the "It's Cool to Fly American" program, which gives young travelers on the spectrum flying American Airlines from nine U.S. airports a simulated travel experience from check-in to boarding. Both programs are offered only once a year.

In addition, some airports are devoting space to make those on the spectrum more comfortable while waiting for their actual flight, writes Judy Antell in her article, "Traveling with Children on the Autism Spectrum":

> In July 2019, Pittsburgh International Airport opened a 1,500 square-foot sensory friendly space in Concourse A, making it—along with Myrtle Beach, SC—some of the first airports in the country to provide such facilities. Pittsburgh, however, was the first to include a real plane, jetway, and multiple rooms for passengers on the spectrum to engage with. The idea came from an airport employee, Jason Rudge, whose four-year-old son Presley has autism. It occurred to Rudge that if the airport had a sensory friendly room similar to the facility at his son's preschool, it would make it easier for families with autistic children to

travel. American Airlines contributed the plane cabin so that passengers on the spectrum who had never flown would know what to expect.[2]

However, if you are not using an airport or airline that offers these programs, Operation Autism suggests contacting your departure airport to

> explain your child's diagnosis and request a time to do an on-site "dry run" of what your child should expect. The airport may allow you to come during a time of day when it is less crowded and let you walk through exactly what will happen on the day you leave, even allowing you to walk your child through the security gate. Take pictures while you are there and add them to your Social Story.[3]

These flight rehearsal efforts are important, says Carl W. Johnson of Adventures in Golf (incorporated as Hamill Travel Services, Inc.) in Amherst, New Hampshire, because

> since every traveler is unique, the preview will let parents know how the traveler handles possible crowds, security lines, fluorescent lights, loud noises, etc. Many people on the spectrum can have extreme difficulty with these circumstances. . . . Rehearsals allow your family or caregiver to assess how the process went for the person on the spectrum and figure out what adjustments need to be made when you do fly. For example, knowing the person's tolerance to various noises might mean noise-cancelling headphones will be necessary, or that they should wear sunglasses indoors if the lighting is too bright. Knowing these things in advance can truly make or break your travel experience. Flying is stressful; removing every possible obstacle can most often offer the best results.

And if you want to take role play one step further, Marian Speid suggests taking your child on a short flight to start. "This will help you both become acquainted with the realistic expectations of spending a longer time in the air."

BOOKING THE VACATION—THE TIME TO RESEARCH AND MAKE YOUR NEEDS KNOWN

Travel professionals and families on the spectrum agree that you should make your needs known well in advance of departure, such as when you initially reserve the vacation. If you book through a CATP, they will take care of this for you, says Sarah Salter, Travelmation dba Sarah Travels Happy in Fayetteville, Georgia. They "can help you order special meals, set up a TSA Cares liaison [more later in this chapter], [teach you] how to pack to avoid excess baggage fees, and even arrange to have a nurse or caregiver accompany you through security as a nonticketed aide."

Naturally, safety comes first when flying—be the child neurodiverse or neurotypical—so even though airlines don't insist on a separate seat for a child under two, it's the most prudent purchase you can make. William J. McGee, author of *Attention All Passengers* and an airline passenger advocate, chaired a subcommittee for the U.S. Department of Transportation in 2010 that examined child restraints onboard commercial flights. He says, "What the Federal Aviation Administration and most airlines don't make clear to parents is that flying with a 'lap child' can be dangerous and even fatal for babies under the age of two during turbulence or an emergency. When I gathered the nation's top safety experts in one room, I asked: 'Is there ever a time when a lap child is as safe or safer than a child seat?' And the unanimous answer was never, so there's no debate about the safety, only about the economics of purchasing another airline seat for a baby. If you dig hard enough on its website, you'll see the FAA acknowledges this."

If you opt to take care of the arrangements on your own, rather than work through a CATP, keep the following tips in mind:

When making the reservation, Marian Speid recommends booking a direct flight whenever possible. "It will ease some of the stress you'd encounter [by taking two or more connecting flights], such as hassles in the aisles and ears popping." Chandra Nims Brown of Cruise Planners—Global Travel Pros in Williamsburg, Virginia, advises traveling during non-peak times "to minimize noise and distractions."

"Consider the seating chart of the plane to accommodate your needs," says Marian Speid.

To that end, Holly Roberts of Frosch Travel—The Travel Center in Hanover, Massachusetts, instructs families on the spectrum to call the airline and request pre-boarding, note in the reservation that you are traveling with a special needs child, and ask what seats would be best for your party. "It is suggested that the forward part of the cabin is best since it is not as loud as the aft," she says. "Sitting up front also prevents children from seeing the large number of passengers behind them," adds Olivia McKerley of Elite Events & Tickets in Grovetown, Georgia.

Operation Autism offers a third benefit: "The front [row] will give you a little extra leg room and will likely put you close to a restroom. If looking out the window would be interesting for your child, request a window seat. On the other hand, if your child may become anxious at the thought of being in the air, request an aisle seat or an inside seat. If the flight isn't full, request a row to yourself," they say.[4]

"If your children tend to kick seat backs, consider an aisle or bulkhead seat for extra legroom," suggests Angela Romans of Rejuvia Travel in South Bend, Indiana. "Let flight attendants know at the time of boarding about your child's needs/habits before a situation occurs (such as constantly pressing call lights, etc.). Lastly, always try to have a plan in place for the potential of delayed flights," she says.

Lisa Bertuccio adds:

> If you are not using a CATP to book your flights or are using a travel agent who is not certified, you can request that the agent enter the special service request code DPNA (Disabled Passenger with Intellectual or Developmental Disability Needing Assistance) when booking, along with an explanation of the additional support you may need when boarding, during the flight itself, and when exiting the airplane. This all goes in the text box usually labeled "additional information." You can enter it yourself if you are booking your own flights.

"That DPNA Special Service Request (SSR) code may allow you additional accommodations for your flight such as priority boarding, next seat kept available, priority rebooking if a flight is cancelled, and other potential inflight accommodations," adds Jennifer Hardy of Cruise Planners in Kent, Washington. She continues:

> Proof of disability is not required, however, the services offered will depend on your specific needs. Bring along a pair of noise-cancelling headphones. The hum of the plane can bother even the most non-sensitive traveler. And always book a standard class or higher. Never book the Basic Economy class. As a family, you will want to select seats together in an area of the plane where you will feel most comfortable. Basic class seats are assigned for you and do not guarantee your family will sit together. I recommend this to all families I work with, regardless of capability.

Along with the special service request, you should also call about ten days before your flight and make sure that the airline knows that you have a passenger who is on the spectrum, suggests Tara Woodbury of Escape into Travel in Leland, North Carolina:

> Ask any questions you may have. Some airlines have someone assigned at the airport to help you with getting through security and to the gate. About seventy-two hours before your flight, call TSA Cares (855-787-2227). You can explain any concerns that you have about your family's needs. They will note that you are traveling with someone who is on the spectrum and will let you know what they can and cannot offer. Different airports have different levels of service: you may be met at the front of security by a TSA supervisor to help get you through security; you may receive a phone call before your flight; or you may be asking for help when you get there. But calling gives you the best advantage. (More on TSA Cares later in this chapter.)

If your itinerary includes a layover, Operation Autism advises researching the intermediary airport in advance to see if there is a child's indoor play area. "More and more airports are offering these. Take advantage of the play area, if available, and let your child run around, climb, play, and slide to expend some pent-up energy and be ready to sit for the next leg of your trip." And if your child is a picky eater or has allergies, Operation Autism suggests contacting the airline in advance to ask if they offer special meals or have an alternate meal option. "For example, you may find the vegetarian meal provides more food that your child will eat than the regular meal. Do not forget to plan for favorite snacks ahead of time as well."[5]

PACKING FOR THE FLIGHT

Your carry-on bag will likely look a little different than for children who aren't anxious, inflexible, or neurodiverse. Marian Speid recommends packing a "go-to" bag with pre-established calming tools, such as noise-cancelling earmuffs, a hat, headphones with a game or toy, fidget toys like Silly Putty or a worry stone (a smooth, polished gemstone, usually in the shape of an oval with a thumb-sized indentation, used for relaxation or anxiety relief), a favorite toy or stuffed animal, chewing gum, a weighted blanket and/or Lap Pad, bottled water, and healthy snacks.

Jennifer Hardy adds, "You may wish to bring your car seat onboard the aircraft if your child struggles with staying buckled or with elopement [escape]. Be sure to check with the airline first to ensure your car seat meets their requirements."

DAY OF FLIGHT

"Plan to arrive at the airport with plenty of time to spare for unexpected situations," says Hardy:

> You will need at least two hours for a domestic flight and at least three hours if you're traveling internationally, an important tip for all families. If you visit a participating airport, be sure to request a Sunflower lanyard. This lanyard discreetly identifies the wearer as someone with an invisible disability. In the event of any emergency or other similar situation, airport employees are trained to look for the lanyard to better recognize when an individual may need extra assistance. It's currently available in Seattle, JFK Terminal Four, and major airports in the UK. The program is also beginning to receive recognition in Australia, New Zealand, Ireland, and for railways and supermarkets in the UK. Proof of disability is not required.

ONCE IN THE TERMINAL

Don't be afraid to ask for more time or special accommodations, urges Jason Adlowitz, a parent from N. Chili, New York. "On Southwest's direct flights to Orlando, you just need to ask an agent at the terminal for a 'Blue Sleeve' [for your boarding pass]. This usually allows the child and both parents to preboard first so they can pick out their seats of choice, though sometimes, they only allow Diana [my wife] to board early with Bailey. Still, this process has helped us out many times."

This is exactly the "divide and conquer" strategy Tara Woodbury suggests to her clients who have a child prone to running away or who will be upset by all the noise. "One of you can talk to the ticket agent or get your boarding passes at the kiosk if needed, while the others keep your child safe," she says.

Regarding elopement, Michelle Zeihr, a parent from Barrie, Ontario, Canada, shares: "We kept our stroller with us (later a jogging stroller as she aged) all the way up until we boarded the plane. It [made it] easier to keep track of a child who may wander off without thinking twice."

Other terminal tips include:

- Check in online [to avoid airport lines] and use the restrooms in the main terminal. They are typically less crowded than the ones [by] the gates. – Sally J. Sutter, Mermaid Travel, Canal Winchester, Ohio
- To keep your child occupied, people watch. Play "Guess Our Plane." Offer positive behavioral reinforcement. – Sarah Salter, Travelmation dba Sarah Travels Happy in Fayetteville, Georgia
- All children need to have things to do in the terminal or while on the plane. Whether it's an electronic tablet, books, or small toys, they can make a tremendous difference. In the terminal, in the case of my grandson, we take lots of walks and bring things to eat before we get on the flight. – Belinda S. Stull, Travel Leaders/All About Travel, LLC, Hagerstown, Maryland
- Be sure to reconfirm your seat [assignment] at the airport prior to boarding to avoid any last-minute changes or issues. – Marian Speid, Dream Vacations in Fair Lawn, New Jersey

GOING THROUGH SECURITY

Check your children's backpacks before putting them through the X-ray machine, says Jason Adlowitz. He relates the following story:

Upon arrival on one of our trips, Bailey, who was around ten years old, had a bottle of chocolate milk with him. Diana, my wife, asked if he wanted to keep the milk to take back to the room or throw it out. Bailey decided to keep it. We arrived at our room and Diana asked Bailey if he'd finished the milk or thrown it away. Bailey told Diana, "No worries, Mom, I threw it out." So pure innocence, no worries, right?

Well, at the end of the vacation when we boarded the Disney charter bus to take us back to the airport, Diana turned to me and said, "I think someone vomited on this bus." When we got off at the Orlando airport, Diana continued, speculating, "I think either Bailey or my bags touched vomit somewhere." I was still hazy about what was going on.

We lined up for TSA PreCheck and loaded our bags onto the conveyor belt when one TSA agent turned to the other and asked loudly, "Whose bag is this?" Bailey was like a stone statue at this point, and of course, it was his bag; the milk from arrival was still inside.

Diana explained the situation and [our son's diagnosis] and assured the agents it wasn't intentional. Luckily, we worked it out and didn't lose our precheck. Once we did get to the gate, Diana felt obligated to see what we could salvage from the rotten milk bag, though after some scrubbing, we decided to throw it and all its contents away.

More on security from Tara Woodbury:

If your child is twelve and under, they don't need to take off their shoes. If having your child walk through the X-ray machine will be an issue, you can carry them. Security can hand scan them quickly and test your (the accompanying parent's) hands and generally pat you down. They will require the other parent to walk through the scanner. In most situations, they will not pat down your child. If they insist on patting down a child with autism and sensory issues, ask to speak to someone higher up.

REGARDING A PAT-DOWN

Autism Speaks gave permission to reprint the following checklist for parents or guardians of children with disabilities regarding the prospect of a pat-down:

- Inform the security officer if the child has any special needs or medical devices.
- Inform the security officer if you think the child may become upset during the screening process because of their disability.
- Offer suggestions on how to best accomplish the screening to minimize any confusion or outburst for the child.
- Ask the security officer for assistance during the process by helping you put your and the child's carry-on items on the X-ray belt.
- Know that at no time during the screening process will you be separated from your child.
- Know that if a private screening is required, you should escort and remain with your child during the private screening process.
- Tell the security officer what your child's abilities are. For example, can the child stand slightly away from equipment to be hand-wanded, walk through the metal detector, or does he/she need to be carried through the metal detector by the parent/guardian?

- Know that at no time should the security officer remove your child from his/her mobility aid (wheelchair or scooter). You are responsible for removing your child from his/her equipment at your discretion to accomplish screening.
- Know that if your child is unable to walk or stand, the security officer will conduct a pat-down search of your child while he/she remains in their mobility aid, as well as a visual and physical inspection of their equipment.[6]

TSA CARES AND TSA PRECHECK

Autism issues, such as non-verbal presentation, dietary requirements, and prescription medicines, can all extend the amount of time it takes to be screened at the airport. Plus, a child's difficulties with sensory issues and being around crowds can make going through those security checks even more challenging. So explains Jennifer Hardy, which is why she points her clients to the TSA Cares and TSA PreCheck programs she uses herself when traveling with her children.

TSA Cares (855-787-2227) is a free helpline offered by the TSA that provides travelers with disabilities, medical conditions, and other special circumstances some additional assistance during the security screening process. Hardy explains:

> For those who struggle with claustrophobia, size of crowds, visual or auditory sensory difficulties, or elopement, the TSA Cares program can provide you with a Passenger Support Specialist (PSS) or Supervisor to assist you with the screening process at domestic airports. The level of support will vary, but may include bypassing the queue, assistance with loading items for X-ray screening, gathering items after X-ray screening, allowing you to carry a child through the metal detector, not requiring verbal responses, and even assisting you with navigating the airport and getting to your gate post-security. Children twelve and younger will not have the unwieldy process of removing shoes and jackets the way adults do. Proof of disability is not required, however, the services offered will depend on your specific needs. It is important to share with the TSA where your child may have difficulty so that they can best assist.

Meanwhile, TSA PreCheck is a program that automatically provides those who register (with or without disabilities) and who are accepted with an expedited and less cumbersome screening process, subject to availability. It is open to anyone, but everyone in your travel party must be registered to utilize the benefits. A background check and fee apply, and membership remains valid for five years. Kids under twelve can access without a membership if everyone over age twelve in their family has their own pass. You can enroll online at https://universalenroll.dhs.gov/workflows?servicecode =11115V&service=pre-enroll.[7]

Sound similar? Hardy explains where the programs differ:

> TSA PreCheck is all about speeding through—like a paid version of a FastPass (or the new Genie+ program at Disney). It really just pre-clears you with a background check so you don't have to take all of your clothing off. Unusual items (like medical supplies) will still require extra screening time.
>
> Meanwhile, TSA Cares is less about speeding through the line (you typically have to wait around for a PSS to meet you before you can "cut" the line), and more about preparing agents for your individual needs—a child who is non-verbal being unable to answer questions, for example. Agents are trained to search for kids being trafficked by air, and an adult answering for an older child is an unusual behavior and a red flag. Using TSA Cares is a preemptive preparation for the agents so that they know it isn't . . . something to pursue further.
>
> Another example: an older child who is fed by formula. Notifying TSA Cares allows them to prepare the appropriate response to you carrying liquid formula through security when you don't have an infant to make it an obvious need. So, it isn't necessarily cutting down on traditional screening times as much as cutting down on the potential extended screening time that would otherwise occur without advance notice.
>
> If you call but don't [need] assistance it really only promotes simple awareness, which may or may not help you as you go through the security process. The only way it truly helps is if the PSS physically meets you and helps you through. Kind of like one of Disney's VIP tour guides: they gather all the information upfront and help you skip over the parts of the screening that would be problematic by taking care of those elements on your behalf without adding time to your journey.

Neither program is a guarantee, warns Hardy. "PreCheck is not offered at all airports, and [while] TSA Cares' advanced notice can be shared with local airports, the support services may not be available [there]," she explains, adding that while they still accept information via phone, TSA now offers an online application at https://www.tsa.gov/contact-center/form/cares.

Hardy's own personal experience with TSA Cares is one reason she counsels her clients to take advantage of the program:

> I am sure people think we make up our children's diagnoses to "skip the line" because much of the time they can seem neurotypical. But on one trip, a TSA officer was unable to meet us right away and the wait to go through security was less than ten minutes. We decided to go through the regular line rather than wait longer for an officer. It was a big mistake. The kids, who are completely used to the security process and had prepared by using their Social Stories and YouTube videos, started to melt down within a minute of getting in line. By the time we reached the X-ray and metal detectors, one of our children was bawling and screaming repeatedly about his environment being too loud and too bright and another was frozen stiff, unable to move or speak. We learned our lesson, and now we steer clear of the regular line and wait for a TSA officer, even if the wait will be longer.
>
> Another [time], we were returning home from a family vacation and had been assigned two TSA officers to help us navigate security. Although they had assisted us from start to finish through the security process itself, our kids started to show small signs of breaking down as we were collecting our items and putting our shoes back on. The TSA officers recognized that two of our kids were approaching a meltdown and stuck around, offering to escort us through the terminal until we could get everyone settled in at the gate.

Starr Wlodarski, of Cruise Planners—Unique Family Adventures in Perrysburg, Ohio, also advises her clients to take advantage of TSA Cares, but, she says, if traveling more often and internationally, go for Global Entry. Global Entry is a U.S. Customs and Border Protection program that allows expedited clearance for pre-approved, low-risk travelers upon arrival in the United States. Members enter the United States through automatic kiosks at select airports. Read more at the U.S. Customs and Border Protection website at https://www.cbp.gov/travel/trusted-traveler-programs/global-entry.[8]

ONCE YOU ARRIVE AT YOUR GATE

There are pros and cons as to when you board the aircraft, and it's really all about how well you know your child.

Tara Woodbury says to consult early with the gate agent, advising them of your names, the name of your child on the spectrum, and your unique needs:

> Explain to them that you will need early boarding or to board last, depending on which you think will work best for your family. Ask them to let the airplane personnel know you would like to get off the plane first with your child or will wait until the rest of the passengers have left (again depending on what suits your family best). They should be able to see a note regarding your initial request for special assistance on your child's boarding pass and/or in their computer so seldom will they deny you. And, just like anyone else traveling with kids, make sure they use the restroom before boarding.

"If your child boards early and then grows anxious while waiting for the rest of the passengers to take their seats, it might be best to board last," says Michelle St. Pierre of Modern Travel Professionals in Yardley, Pennsylvania. "You can walk around instead of waiting but make sure the gate agent is aware, so they don't shut the doors on you."

Conversely, Operation Autism suggests that for some children, boarding early might be preferable because it gives the child with autism more time to reach their seat and get comfortable. "This will also give you extra time to briefly talk to the airline attendant(s) about your child's special needs. Give them an information card, if necessary,"[9] Taylor Covington from *The Zebra* elaborates. "If you inform the security officer and flight attendants of your child's needs before an incident arises . . . this will help them respond more calmly to your child's issues if they develop problems on the flight."

KEEPING BUSY ABOARD THE FLIGHT

Once you're on board, there's still much to deal with until you land, and all of it is unfamiliar to a child on the spectrum new to air travel.

For example, Operation Autism points out that children with autism often have difficulty coping with the "popping" of their ears as the airplane ascends and descends. "Many children do not understand this phenomenon. Give your child gum, hard candy, chew toys, or a drink during this portion of the plane ride; it encourages swallowing and reduces the effects of their ears popping."[10]

Then there's the emergency exit plan explanation. Marian Speid urges parents to "limit distractions for your child during this procedure. You may want to wait to give them their toys as a reward for paying attention to this pre-flight training. If your child likes to press the call buttons, let the flight attendants know ahead of time. You can provide an alternative object with buttons to push instead."

"If your child will wear headphones, listening to some favorite music or videos will be a big help and can keep the noise down on the flight. [Plus], anytime you can surround your child with family and give them a good sense of security, it helps calm everyone," says Belinda S. Stull.

In-flight dining has been an ongoing issue for parent Michelle Zeihr of Barrie, Ontario, Canada:

> Most of the time, the [airline] would allow us to bring sandwiches, snacks, juice boxes from home on board. This was great because some flights offered foods that Thessaly just wouldn't even touch. For example, Icelandair included kids' meals. One of them was some chilled noodles with veggies and chicken. There was no way we could have gotten her to eat any of it. Luckily by that point, we knew to bring things. That was true for things to keep her busy before and after the meal as well. We had a tablet with movies downloaded on it, as well as a portable DVD player. We brought coloring books and crayons, anything to try and keep her occupied.

Tara Woodbury offers these two pieces of advice to encourage appropriate onboard behavior:

> If your child is motivated by "prizes," you may want to have something small they will enjoy for each phase of the airplane ride—boarding, taking off, cruising, landing, waiting to deplane, and leaving the aircraft. Remember that story board? Bring it along. It can serve as a physical reminder of what to expect.

Secondly, if your child likes to know how long something will take, a small digital clock or watch counting down the duration of the flight may help. Stick anything else you might need in a hurry in the seatback pocket in front of you.

Most of these tips will help neurotypical children as well. No children are known for their patience and traveling disrupts both nap time and schedules.

DEALING WITH UNEMPATHETIC ONLOOKERS

The advice in this section could really apply to any chapter of this book. The unfamiliar makes people nervous, and a child with special needs can often attract unwanted attention, especially at an airport or on a flight, including hurtful comments and unpleasant looks. How should a parent handle such negative energy and shield their child from it?

And the looks don't always come from other travelers, says Sarah Salter, who believes that in some instances, supplier training could be better:

> The onus should be on the destination to want to work with people with various abilities of all types and ages. For example, my son uses a "wheelchair as a stroller" tag at Disney for when he is getting his feeding tube. Even though this accommodation is in place, we still have interactions with cast members and/or other guests that are poorly trained. This was especially the case when we used a Keenz [stroller] wagon. Even though we followed procedure and had the appropriate tags from guest relations, we got a lot of "Sure wish I could ride around like that" or "How lucky are you to rest in a wagon," etc. We also got a lot of side-eye, and "helpful" people (other guests) pointing out where the stroller parking was.

Parents need to accept that addressing their children's needs attracts far less attention than their children losing it in public, says New York–based clinical psychologist Dr. Ellen Littman. "As the parents internalize the following concept, they can explain to their children: Everybody is different, with unique tastes, interests, energy levels, and frustration levels, and we do the things that work best for our family."

For some I spoke with, like parent Michelle Zeihr, it's not an issue. "So far, we have not had anyone give us odd looks or stares. . . . Thessaly has been aware of her autism since she was very little; I've taught her to embrace it and let it empower her. Everyone is different and those differences make us who we are. Thessaly will often say to someone who does meet her: 'I have autism and when I'm really excited, I flap my arms.' She says it with a big smile on her face," Zeihr shares.

Cathy Winter of EnVision Travel in Denver, North Carolina, has a son in a wheelchair. She says he is "always getting 'the look' or being talked to like he is 'not there.' I encourage him to take control of the situation before it happens by saying, 'Hi, how are you today?' It makes the 'looker' realize that my son is what you call 'normal' but just in a wheelchair. I always advise people to talk directly to the person who may be challenged or disabled regardless of their cognitive level."

Kristen Chambliss says her reaction to any rudeness depends on the situation. "Usually, I remove my daughter and take her somewhere to calm down, without feeling the need to explain her actions to others. Our daughter does great, though, so this hasn't been an issue for us since she was three. She's almost eight now. We have a list of coping skills she can read and choose from to calm down (e.g., getting a drink of water, playing with a pop-it, taking some deep breaths, etc.)."

For others, like Jennifer Hardy, a CATP and also the mother of four special needs children, it's been more difficult. "It has personally taken me a long time to ignore when other people look at my family funny or say something insensitive," she says. "The most important thing is to focus on your child and their immediate needs. Unless someone is offering to help you, tune everyone else out. You do not owe anyone an explanation, and no one deserves to come between you and your child in a time of crisis. I don't even pay attention to odd looks anymore because the overreaction isn't worth the attention."

"Unfortunately, we will always encounter some rude or insensitive people in our travels," says Sandra Chu of The Enchanted Traveler in Rockford, Illinois:

> It's important not to let that interfere with enjoying the vacation that you're entitled to. Hopefully, for every one of those insensitive people, there are far more people who are kind and accommodating, as our societal awareness of sensory issues increases. It can be helpful to take advantage of resources like the free IBCCES Accessibility Card (IAC) which helps individuals who may need extra assistance at theme parks and other attractions. This can help to reduce the risk of meltdowns, get the assistance needed

efficiently, and reduce the risk of misunderstanding the nature of a child's response at a stressful time. (More on the IAC in chapter 14.)

"Always choose your child over the shaming and condescending stares of the public," says Brooklyn parent Delight Dolcine. "If you think it's hard for you or the people complaining, imagine what your child is experiencing." Her strategy?

> You should explain proper behavior and protocol in advance of the actual flight. . . . We practice [working with] physical boundaries since Xenia has trouble giving people space. She walks on the back of their heels and bumps into people or will walk right up to someone and clap or laugh hysterically in their face. Some people are intimidated by these behaviors. I've had to give a lot of prompts about [this]. Giving people space is also an issue on the plane. She would peek in the crack of the seats in front or behind us and laugh, which freaks people out. So, etiquette [lessons are a must], mainly for the terminal and on the plane.

"Remember that other passengers will not know your child has autism," says Taylor Covington. "If your child is not having trouble, explain to [those] around you about your child's needs, so they will be more understanding if a problem [does] arise. This will elicit compassion, not judgment, from your fellow passengers."[11]

Lisa Bertuccio explains to clients that she's not trained to offer advice on a professional level like a social worker or psychologist would,

> but as a mom . . . my main concern wouldn't be what other people around me were thinking. It would be keeping my child safe from hurting themselves in the event they were having a meltdown. I think if the person was still there after my child calmed down, I would probably approach the person and explain what they had witnessed was just a meltdown because he was non-verbal and couldn't express himself the way we would or that he is sensitive to certain sounds or whatever the trigger was for the episode. Oftentimes people are just curious and if you can help educate and help foster an understanding or compassion for a situation, people usually change their future reaction to it. There are also many products that you can find on Etsy that can help people identify that the child has ASD. These include buttons and lanyards with tags that say that the child is nonverbal, ones that ask for people to be

patient [with children that] have autism. Medical alert bracelets can be customized to the child's specific disability.

"Our family has found it best to simply avoid the people who like to stare and make funny faces," says Carl W. Johnson:

> There is nothing you can say or do that is going to make them stop looking. Most of the time, we find that people are very understanding that we have a unique situation. For one of our children, it is readily apparent that they struggle generally; most people pick up on that notion quickly and will actually give us a brief smile. Or even sometimes, we find other people being helpful—moving out of the way near security to give us more room, or a simple face or hand gesture to let us know they are aware that we need a bit of extra time or whatever we might need at that point.

"I always let my clients know that this is bound to happen and just to shake it off," says Sarah Marshall of TravelAble in Midlothian, Virginia:

> But it is one of the reasons why Certified Autism [training] programs are important. These children should be given the space to be themselves without needing to explain themselves at every turn. Usually, we pre-plan things so that they are seated in quiet areas, take private tours, and are given opportunities to shine without needing to explain to all the mean people in the world.

One mom, who preferred to remain anonymous, describes how she handles less-than-empathetic onlookers:

> On board our first flight when he was five, my son kept his voice low, did not demand anything, did not want to walk back and forth, did not cry, or get upset during six long hours strapped into a chair. All he did—when not eating or watching a movie—was pop up and down in his seat or lean on his fold-down tray to draw. As far as I was concerned, he was gold, even though the woman sitting in front of him didn't agree because her seatback would move as he drew. I assumed she'd never been a child herself and had sprung fully formed from her father's head. So, my advice is to ignore assholes who don't realize that children are not adults.

ALLEVIATING MOTION SICKNESS

Motion sickness can occur on any mode of transportation, and some of the comments in this section reflect that. Unfortunately, spectrum children may be more susceptible than neurotypical kids because "the one thing that is often recommended to make [their] travel easier—distractions with movies etc.—can actually make motion sickness worse," warns Tara Woodbury. Medication isn't an automatic solution either, especially if the child takes drugs for their special needs, because of potential contraindications (negative interactions between meds). For these and other ethical reasons, all CATPs surveyed said they would not presume to recommend any prescription meds to their travel clients. That being said, here are some holistic and over-the-counter advice they offered.

"We try to have a focal point for [my daughter] to stare at. Whether it's the seat in front of her on a plane or in a car, we encourage her to look at it when she starts to feel woozy," says Woodbury. "Looking outside—though it's the standard advice these days—often makes it worse," she says.

"I use essential oils, especially peppermint, because the smell helps calm motion sickness. Fresh air when possible is always a good [idea]," says Cathy Winter.

Lisa Bertuccio has a similar strategy for her child:

> For car rides, I make sure he is not looking down for an extended period of time. That means things like tablets, books, etc. are a big no-no on car rides. Instead, I have him watch movies on the TV screen located on the headrest. I also give him gum to chew or have him suck on ginger lollipops, which seems to help with any nausea he may have. Another remedy I use for my son are Sea-bands (reusable, inexpensive, anti-nausea acupressure wristbands that come in both child and adult sizes). They sell generic ones that . . . have a button-like thing sewn into the band that applies a gentle pressure to a specific area of the wrist which combats motion sickness. I really didn't think they would work but they really do! I even used them while I was pregnant, and it was the only thing that helped with my morning sickness.

Sandra Chu reminds parents that if they choose basic Dramamine or ginger in various forms (capsules, ginger ale, drops, chews, gum) to prevent or lessen motion sickness, their children must take it an hour before travel.

Carl W. Johnson warns that while wristband products are helpful, some children on the spectrum might not like having something on their wrists. "If a child or family member has this issue . . . then it might force you to delay your travel plans until you can test [other] options to see what helps them best. I would suggest speaking with their pediatrician about possible medications to take during the trip."

Tara Woodbury found a non-medical, non-wristband solution for her own daughter, whom she describes as an anxious and somewhat neurodiverse traveler: acupressure. "Put three fingers on the inside of your wrist (or your child's) and your thumb on the other side, and apply light pressure," she recommends.

Meanwhile, Sarah Marshall suggests speaking to a doctor about ordering a special patch if her two go-to solutions—Sea-bands and Dramamine for Kids—aren't effective for a particular child.

Jennifer Hardy looks at this problem from personal experience and mostly from a cruising standpoint:

> I try to avoid medicating myself against motion sickness. Instead, I focus on selecting a stateroom on a lower deck in the midship area or sometimes towards the aft. The higher the deck and the closer to the front of the ship, the more motion someone may feel. If all else fails, I have found eating a tart green apple significantly helps in reducing symptoms. You can request an apple from room service, or they sometimes will have them available at the buffet. And of course, the ship will stock seasickness medications in their gift shops, medical facility, and at Guest Relations.

From pre-flight preparation to dealing with motion sickness and inconsiderate onlookers, the strategies covered in this chapter should get travelers with anxious, inflexible, or neurodiverse children one step closer to their destination. Coming next, I'll cover some alternate forms of transportation, such as cars, buses, trains, and ships.

PARENT SPOTLIGHT—
JASON ADLOWITZ, N. CHILI, NEW YORK

Jason Adlowitz is a parent with one son on the spectrum, diagnosed with Asperger's syndrome and attention-deficit/hyperactivity disorder. He's been traveling with Bailey since the child was five years old, starting small with a trip to visit family in Syracuse and then making a trip or two to Toronto:

> Our first actual [vacation] was back in January of 2013 when we decided to travel with our in-laws to Walt Disney World. Many of our concerns arose in that first trip: weather; overall adjusting; sensory issues in relation to food, clothes, sweating, and [meeting new] people. Would Bailey get overstimulated or not have enough input? How would we handle meltdowns when a situation became overwhelming? If we flew, how would [going through] security, waiting, departing, and arriving affect him? Bathroom concerns. You name it, we had reservations about it. And yet, thanks to our agent Nicole Thibault, we've become a Disney World family because we find that Disney creates a safe environment for everyone, especially children on the spectrum.

Adlowitz says the biggest mistake would be for the parents of a neurodiverse child to automatically decide that travel would be impossible or that you can never go on vacation again. Instead, he advises parents to question what resources are available and "explain [the trip in advance to their child] via videos or pictures through the internet. Our experience is that the more information a child has available, the easier the transition from one activity to another. Giving choices has also worked well for us."

CHAPTER 8 ACTION STEPS

1. Introduce the idea of airline travel long before the flight through Social Stories, picture books, role play, visits to programs like Wings for Autism, or just an advance visit to the airport with programs that a particular airline may provide. Take a short flight in advance of a longer one as a test run, if possible.
2. Seek out sensory spaces in airports that offer it.
3. When making your reservations, use a Certified Autism Travel Professional, or—if you're booking on your own—be sure to alert the airline to your child's special needs. Look into programs like TSA Cares and TSA PreCheck. Consider which seats might work best for your child and request them in advance. Don't leave seat assignments to the last minute.

4. If your child is a picky eater, bring food along for the flight or arrange for special meals that your child will like. Alert the airline to any allergies.

5. If your journey involves a layover, part of your advance research should include investigating what services are available at the airport of your intermediary stop.

6. Make sure your carry-on luggage includes items that will both distract and calm your child before and during the flight. Include necessary medications. Also consider bringing your car seat if your child struggles with staying buckled or with elopement. Make sure the seat meets airline specifications.

7. Arrive at the airport early and take advantage of any special identification items the airline or airport might offer for special needs families.

8. Don't be afraid to have one parent check in while the other occupies the child's attention. Keep your stroller with you as long as possible.

9. Be aware of your rights for going through security, including pat-down rules.

10. Decide the best time to board based on the needs of your child.

11. Bring gum and hard candies (if safe for your child) to alleviate ear-popping during the flight.

12. Keep the child engaged during the flight, especially during the emergency exit explanation. Provide alternatives to pushing call buttons that can disturb other passengers and flight attendants.

13. Decide in advance how to handle unwanted attention from curious onlookers and possible motion sickness.

9

ROAD TRIPS

Negotiating Car and Bus Travel

"We attempted two car trips to Disney World, both with their own set of interesting surprises. We had a lot of issues with bathroom breaks, with Bailey not wanting to go, then getting back into the car and announcing that he actually did need to go. There were traffic issues: if a toy/book/LEGO® fell out of reach, meltdowns would occur, and redirecting while in motion was sometimes tough." – Jason Adlowitz, parent, N. Chili, New York

"On car and bus transfers, like on airplanes, we have the same issue of Xenia laughing loudly and inappropriately, clapping in people's faces, refusing to stay seated, or blasting her YouTube videos on her phone. She refuses headphones and is extremely sensitive on her earlobes, so I lost that battle a long time ago. But I say things like 'Quiet mouth, Xenia, let's be respectful,' or 'Safe hands,' so she doesn't swing her hands or touch people unwarranted." – Delight Dolcine, parent, Brooklyn, New York

Cars and buses can be wonderful alternatives to air travel. Most families must already realize this since in recent years, nearly 80 percent of family vacations have involved spending some time on the road.[1] "Car travel is more convenient than bus or train travel since you are in control of your stops," advises Holly Roberts of Frosch Travel—The Travel Center in Hanover, Massachusetts. "You can pull over when your child needs a break and have some downtime outside of the car."

And yet, hours trapped in the vehicle, especially strapped into a car seat, can be hard on any child.

Even if you're flying or sailing to a destination, cars and buses may still be part of any trip that requires transfers from the airport or dock to a resort or hotel. Some of the helpful tips that follow might echo what was recommended for air travel, but I've included them again for those who may be reading only certain sections of this guide to prepare for a particular vacation. As always, start small before embarking on a major journey.

IN GENERAL

"While most people don't think about bus travel, it may be a good option, as [buses] usually make frequent bathroom/stretch your legs stops to break things up for your child," says Jonathan Haraty of Adventure Luxury Travel in Whitman, Massachusetts. "If you're using a van or a bus for transfers to your resort location, they may be crowded at times, but the rides are usually of short duration."

However, for some children on the spectrum, tight spaces, crowds, moving buses or trains, or even a short hop on an airport tram between the terminal and baggage claim *can* cause significant issues and you should be prepared for them, warns Carl W. Johnson of Adventures in Golf (incorporated as Hamill Travel Services, Inc.) in Amherst, New Hampshire:

> [If you can't] avoid them entirely, implement whatever measures you can to help this process. One of our children has a very tough time with airport trams, so we do everything in our power to make it as fun as possible. We ensure they have their iPhone and headset at the ready! We then "talk it up" as the doors are closing, saying things like, "The train is very short—you can even see the other side," or "When we get into the terminal area, let's get an ice cream" (or another favorite treat). Simply giving the person something positive to look forward to (in the *very near* future) can have a massively positive impact [on] the stressful moment or travel mode which is causing undue stress.

Talking about what is happening is a good idea when dealing with children of any age, according to Belinda S. Stull of Travel Leaders/All About Travel, LLC, in Hagerstown, Maryland. "Car travel is a little more convenient [and there] are maps and apps available that show playgrounds near the interstates for a longer break. Bringing along small toys, electronics, and favorite items will help make the car or bus trip bearable."

PRE-TRIP PREPARATION

Parents and Certified Autism Travel Professionals agree that when it comes to road trips, good preparation is half the battle. If you're driving, that means prepping your vehicle ahead of time.

"To avoid unexpected surprises en route, have any [car] maintenance completed before leaving home," advises Angela Romans of Rejuvia Travel in South Bend, Indiana. That might include "having your car seats and booster seats checked professionally before you hit the road to ensure they are properly installed," says Taylor Covington of *The Zebra*. "Poorly installed car seats are not only a safety risk, but they are also uncomfortable, and that can lead to the wrong type of sensory input for your child."[2]

If you're planning a long car trip, be aware that some children are sensitive to the interior noise. Tires can be one issue; the brand of car can be another. One of Car Talk's blog transcripts (www.cartalk.com /content/my-youngest-son-has-been-diagnosed-autism-one) contains the hosts' recommendations.[3]

If renting a car at your destination, bring your own car seats, says Starr Wlodarski of Cruise Planners—Unique Family Adventures in Perrysburg, Ohio. Seconding that opinion, Jenna Murrell, a digital marketing specialist for the safety advocacy group Safer America, explains that while some car rental companies will provide free seats if you redeem reward points or are a member of AAA, it might be a safer bet to bring your own.

In a 2016 blog post on *Safe Ride 4 Kids*, Murrell recounts how *Consumer Reports* sent out child passenger safety technicians to two large rental car companies to analyze the car seats available for rent that day:

> One agency . . . kept their seats in clear plastic bags with owner's manuals zip-tied to the covers. The second stored their seats outside in the parking lot shed, [where they] actually appeared to be thrown rather than placed down. Upon closer inspection, the technicians found almost none came with the owner's manual, while some had missing or broken pieces, infant seat carriers [that were] separated from the base, or were altogether expired.[4]

Her point: safety should always come first and since *Consumer Reports* didn't name either company, should you really take the risk?

Planning the itinerary is equally important for road trips with children on the spectrum, says Jennifer Hardy of Cruise Planners in Kent, Washington:

> Prepare your driving route in advance and be thorough. Be sure to research rest areas and locations to eat. You will also want to allow extra time throughout your journey to account for traffic delays and unexpected stops. Make sure you leave at a time that works best for your child—during the day if they do best when awake, or at night if you prefer your child to sleep while you drive. And of course, review the itinerary with your child and try not to stray from the schedule. Bring your child's favorite foods and toys along for the ride! Always pack more than you think you will need.

Try to plan around non-peak travel hours to avoid traffic and crowds, adds Nicole Graham Wallace of Trendy Travels by Nicole—Dream Vacations in McDonough, Georgia.

Next, for car or bus travel, it's vital to take steps to get the child ready. For some children, it's a matter of stirring up some enthusiasm. "We tell our daughter we are going to get into a new car or bus, emphasizing how exciting this is," says parent Kristen Chambliss of League City, Texas. But other children need more than hype. Sarah Salter of Travelmation dba Sarah Travels Happy in Fayetteville, Georgia, tells parents to go even further, explaining each step and transfer ahead of time at the child's level, "For an older or high-functioning child, a brief verbal reminder may be sufficient. For a younger or lower-functioning child, draft their best friend: 'Monkey is so excited about riding on a bus! He's never been on a bus before!' or 'Monkey is a little scared of the loud engine. Can you show him how to be brave?'"

"Consider creating a custom story with pictures and details of things they will see and experience along the way (or find a suitable Social Story based on your destination). Read the story days or weeks ahead of time," counsels Angela Romans. "Helping them [pick] special sites to visit during the drive is [also] great," adds Olivia McKerley of Elite Events & Tickets in Grovetown, Georgia.

Bus travel also needs some advance preparation. "Choose the seat with more legroom and notify the bus driver or ticket agent of the need to board first and exit first. Pack essentials that will occupy the child or act as a comfort, such as toys, coloring books, snacks, a lovey, and video player with headset. Many of these tips work well for neurotypical children and I have used them with my own children when we travel," says Lisa Bertuccio of Vacation Your World in Otisville, New York.

AIRPORT TRANSFERS

Transfers require a different type of attention, according to experts. "If possible, arrange to rent a car for your family to [travel between] the airport and the hotel or resort," suggests Tara Woodbury of Escape into Travel in Leland, North Carolina:

> Waiting for resort transportation can be time-consuming and hard for most children on the spectrum to understand [considering] the crowds of people and the fact that the same [special needs] accommodations don't exist for buses as they do for airplanes. If you must take a bus or resort transportation, use the same coping techniques as you would on an airplane to help the time pass quickly. Another alternative might be to Uber to your destination from the airport.

Angela Romans also promotes private transfers but if that's not possible, she offers a different suggestion:

> For transfers from the airport to your destination, make sure you know the approximate distance [and communicate that to your child]. Will the driver be making stops at properties along the way, or heading directly to your hotel or resort? In destinations such as Jamaica, the trip from the airport can take roughly an hour and the roads can be long and curvy. The drivers usually stop halfway for guests to take bathroom breaks or pick up sugar cane to help alleviate car/motion sickness.

"Nate is not great at waiting, and we're never certain how he'll act in large crowds," explains parent Michael Sokol of Longmeadow, Massachusetts:

> I like to arrange private transportation whenever possible. Our cruise from New Jersey stops in Florida for the day, and we always head to Disney/Epcot, but we do not take the bus transfer. My travel agent arranges a private car for us. This allows zero wait time getting to Disney, and when we are ready to go, we call him and bam, we go back to the ship. It's faster, no waiting, no loud bus for Nate . . . smooth. In the

Bahamas, Nate loves the Atlantis. It's a great day excursion during our cruise stop. When you get off the ship there are vans, taxis, and buses lined up to take people all over the island. WALK past all of them into town. It's only one block away and once there, you can grab a taxi with no waiting, no sharing—it's faster and cheaper.

WHAT TO PACK

What you bring along on the trip can contribute greatly to its success. So be sure to create a customized travel bag for your child filled with some of their favorite (car-friendly) toys, advises Marian Speid of Dream Vacations in Fair Lawn, New Jersey:

> During the trip, the front-seat passenger can hand these off as needed to occupy and entertain your child. Make a go-bag, filled with necessities and pre-established calming tools. Some examples: hat/sunglasses; headphones with music or games; fidget toys like Silly Putty or a worry stone; Lap Pad; favorite toy/stuffed animal; healthy snacks; chewing gum; clothes for layering; and more.

"Car travel allows you [the extra space] to bring your own blankets and pillows," adds Sarah Salter.

For distraction, Taylor Covington urges parents to pre-load an iPad with old and new games for single- or multiplayers, and kid-friendly television shows and movies.[5] Canadian parent Michelle Zeihr of Barrie, Ontario, brings a tablet, a cellphone, and "an Echo Auto which Thessaly loves—it answers questions, plays songs and games." Just don't forget extra batteries for electronic devices, says Angela Romans.

ALONG THE WAY

The key to a successful road trip, say experts, is to ward off problems before they turn into meltdowns. Romans counsels: "Pay attention to your child's emotions. If they start getting agitated, take an unexpected break. Watch out for motion sickness as children who are neurodiverse or have sensory integration [issues] can get carsick more easily than others. Carry extra clothing and cleaning supplies as an added precaution." A supply of wipes and trash bags wouldn't hurt either, says Sarah Salter. (Read more about handling motion sickness in chapter 8.)

To keep things as routine as possible, try to maintain your regular schedule as best you can, says Taylor Covington. "Do you normally have lunch at noon? Plan your trip so you can keep this time frame. Sensory toys may be helpful at certain periods during your trip. Consult with your child's occupational therapist for specific sensory strategies." Here are her suggestions of some items that can promote calmness and relaxation:

- Vibrating massagers
- Hard candy (if this does not pose a safety hazard for your child)
- Vibrating beanbags or teethers if your child needs additional oral input[6]

Operation Autism offers these additional road trip strategies:

- Use a behind-the-seat organizer to store your child's belongings. If they are within your child's reach, it will enable you to focus on your driving and maintain an organized car.
- During rest stops, allow your child to run around outside if it is safe to do so. If your child needs a visual cue to know when the next stop will be, bring along a timer and set it for sixty minutes. That way, your child knows that when the timer goes off, they will be able to get out of the car. Alternately, you can ask an older child to time the interval on the clock in the car and then pull the car over for a break at the intended time.
- Get a map and draw a line along your route. Show your child each city and state you are passing through. Use the line to show your child where you are "now" and where you need to drive before you get to your final destination.
- Play creative games using the things you see along the road:

  ○ Bingo game: before your trip, prepare a Bingo sheet using pictures or words of things frequently seen on the road.
  ○ Alphabet game: take turns finding things that start with a designated letter until you get through the entire alphabet.

- ○ License plate game: write down all the license plates you see from different states and see how many you can find by the end of your trip.
- ○ See http://www.MomsMinivan.com for more car game ideas that you can modify for your child's age and developmental level.[7]

Finally, Marian Speid says to remember to reinforce and praise appropriate behavior along the way. "Consider bringing a new toy or other incentive for encouragement," she suggests.

Though the challenges are surmountable, as the preceding pages make clear, cars and buses aren't the only forms of ground transportation. In the next chapter, I'll cover train travel, which is a passion of many children with autism spectrum disorder.

PARENT SPOTLIGHT:
KELLIE ENGSTROM, KENT, WASHINGTON

Kellie Engstrom is the parent of three children, with two on the spectrum, one with autism (age thirteen) and the other with sensory processing disorder (age six). She's been traveling with them since the oldest turned three, looking to relax, but her host of apprehensions were anything but calming: the fear of elopement and finding food the kids would eat. In addition, without the sensory room they had at home, and the ability to provide one child with enough garments to accommodate myriad daily clothes changes, she knew going on the road would be challenging.

"I definitely started having more fun when I accepted that this was not going to be relaxing. Instead, looking at it as just a new experience for the family made for a better mindset," Engstrom says.

She notes the traveling-style differences between her neurotypical (NT) child and neurodiverse (ND) children: "It took a little over 24 hours for my ND kids to adapt to the new environment. Also, the amount of time we can spend at one place is less with the ND kids. My NT daughter was much more adaptable. We also try to find ways to give her breaks from her brothers. Having to always flex with what they need gets difficult [for her]."

While Engstrom insists it helps everyone to know what is coming, she tries to keep daily schedules loose. "Breaking up a journey and shortening the length of time spent in the car makes things easier on everything," she says. But sometimes, those breaks can be unexpected:

> On one trip, about an hour into our drive, I realized I'd packed us up a day early. Going home would have resulted in huge meltdowns. I called Jennifer Hardy, my travel counselor, and she found us a hotel halfway down. My thirteen-year-old still likes to tell people how we got an extra day because Mom messed up. On the way back, we drove straight through and that made us realize how much easier that overnight break made it on everyone.

Engstrom also likes to end with a swim for the kids. "It's for sensory input, but it helps them all calm down and sleep better," she says.

Another tweak: "Some of my family need more time to wake up than others. We finally figured out to have the 'awake' kids help get breakfast for the rest of the family."

In the end, she says that what makes travel possible for the family is accepting who her children are. "My kids don't sit still. They like to go. Planning trips that don't involve 'relaxing while looking at the view' helped a lot. Remember: [holidays] come in all shapes and sizes. Don't get wrapped up in the idea of the perfect vacation."

CHAPTER 9 ACTION STEPS

1. Avoid using buses for airport transfers whenever you can. Private transfers may be worth the extra expense in terms of sparing your child any additional stress.
2. If you must take bus/tram transfers, create an excitement around them and link them to favorite fictional characters. Offer the prospect of a reward on the other side.
3. Describe as much of the journey as possible in advance. Use Social Stories when possible.
4. Advance planning for car trips should include vehicle maintenance, making sure car seats are installed correctly, minimizing interior noise, and mapping out the route to determine frequent stops (at playgrounds and rest areas, for example). Schedule the journey to coincide with your child's routine and rhythms, and try to avoid peak travel hours when traffic and crowds could become a problem.
5. Let children suggest parts of the itinerary based on their interests.
6. If you are renting a car, bring your own car seat.
7. For bus travel, arrange for the best seat location for your child. Time your boarding (first, last) to coincide with his or her needs.
8. Whatever your mode of travel, make sure you have adequate supplies of food, comfort items, and electronics with headphones to provide a distraction. Unlike other forms of transport, cars provide the extra space you may need.
9. Have creative "travel games" at the ready.
10. Praise and incentivize appropriate behavior.

10

CHUGGING ALONG

Train Travel

"Train travel is generally a great way to see the United States, especially when traveling to the national parks in the West. You can visit the club car or the observation car with your child on the spectrum and if there are any triggers or spatial issues in either of these environments, you can go right back to your private room." – Jonathan Haraty, Adventure Luxury Travel, Whitman, Massachusetts

For children wary of flying and parents unwilling to cover long distances by car, train travel might be the perfect compromise. One benefit is that unlike traveling by car, parents can devote 100 percent of their attention to their child because no driving is involved. Another plus is that it might be an easier sell, especially since according to developmental pediatrician Amanda Bennett, neurodiverse children harbor a certain fascination for locomotives.

Dr. Bennett explains that

trains have wheels, and this will appeal to those whose sensory interests include watching objects spin. . . . Second, trains can be categorized into different models, types, sizes, etc. For some individuals with ASD, the ability to organize objects into categories is very appealing. . . . Trains also come with schedules. This, too, appeals to many people with ASD and is in line with a need for predictability and the inclination to memorize and recite information. . . . Plus, trains have features that can support an enduring and developing interest across a lifetime. For instance, many children—including children on the spectrum—love the train characters in videos and TV shows such as "Thomas the Tank Engine" and "Chuggington." For older children and adults, model trains can be fun to build and have mechanical features that can be interesting to take apart and reassemble.[1]

The love of trains is so pervasive among children with autism spectrum disorder (ASD) that the transit museum in New York City and many similarly themed museums around the country are working together to jointly promote programs to appeal to this audience and help them to better engage with others.[2] There's even a private Facebook group called Transport Sparks[3] catering to parents of neurodiverse children who are transport fans.

Along with children's fascination with trains creating an eagerness to travel, another bonus is that Amtrak claims to have one of the most generous baggage allowances in the travel industry. Passengers can bring two carry-on items aboard (up to fifty pounds each) and two personal items (up to twenty-five pounds each), for a total of 150 pounds, for free.[4] (Additional bags can be checked in, where permitted.) This allows for more familiar items to be brought along for comfort and security.

Parents can play on their child's enchantment with trains by reminding those on the spectrum that getting there is half the fun, says Michelle St. Pierre of Modern Travel Professionals in Yardley, Pennsylvania:

> This goes for monorails, trains, and buses at your destination as well, so build that into the planning when you talk about your trip. For example: when you fly into Orlando International Airport, you must take the monorail to baggage claim, and then a bus or a taxi to Walt Disney World. When discussing your plans, be [enthusiastic] about the number of "transfers" involved—that it's exciting to try so many forms of travel—otherwise your child might melt down if they don't see Mickey Mouse the second they step off the plane. This is a good tip for all kids, especially first-time travelers.

While children on the spectrum may be pre-sold on the idea of a vacation on the rails, and may, in fact, be soothed by the train's rocking motion, parents and Certified Autism Travel Professionals warn that there are still some precautions families should take to ensure a successful journey. "Each person on the spectrum—and even neurotypical people—have auditory and sensory issues that you must anticipate; they simply cannot be ignored," warns Carl W. Johnson of Adventures in Golf (incorporated as Hamill Travel Services, Inc.) in Amherst, New Hampshire. "Once on the train, it's not stopping until it reaches its next destination. You simply do not want someone having a sensory overload in a location or setting which you cannot control and/or stop completely."

Betty J. Burger, recently retired from BGB Travel and Leisure, LLC, in St. Marys, Georgia, attributes these sensory issues, as well as stress-causing delays, to the uniqueness of train travel which has "sights and sounds not experienced on other modes of transportation." She explains that in America, "rail lines are 'borrowed' by passenger trains; commercial trains [take priority and] can cause delays along the route. Movement may not be as restricted as on an airplane but is still limited. A visit to a passenger terminal in advance of travel [will let] the individual experience the [station] as well as view the trains and hear some of the sounds involved," she recommends.

In fact, train stations can be the destination in themselves, says New York–based clinical psychologist Dr. Ellen Littman:

> Once they get on the train, children usually get bored pretty quickly since they have to stay in their seats and can't explore. [By making the station the destination] children can go to the tracks and examine the outside of the trains, getting a close-up of the wheels, the conductor. Hopefully, this can help parents of [rail fans on the spectrum] avoid paying for a touristy train ride that their children may be less than enthusiastic about.

The success of this tactic will depend on the size of the station in question, as well as the child and his or her sensory issues or other triggers. "Some terminals can be as hectic as an airport—think New York's Grand Central and Penn Stations," warns Sally J. Sutter of Mermaid Travel in Canal Winchester, Ohio.

And it's not only the commotion of the station that can cause issues. London-based parent Rebecca Da Re says that train station challenges for her ten-year-old, neurodiverse son Romeo include ticket barriers (called turnstiles in the United States), which can frighten him, and escalators. "Romeo has issues with spatial awareness so it's often me shadowing him to help him navigate and using the wide entrance barriers with both going through at the same time," she says.

Carie-Louise Gray, from Sussex, United Kingdom, found that her train-loving son Benjamin-Edward, diagnosed with autism and expressive and receptive communication delay, had a huge fear of the color red up until two years ago, "so going on a red train was hard. But now it's his favorite color and he loves the red trains," she reports. Something that helps mollify any issues: chocolate. "One station we go to, Portsmouth Harbour, is near my favorite chocolate shop. . . . He loves their chocolate coins and teddies so he gets some too."

Once you've conquered any issues with the terminal, Sandra Chu of The Enchanted Traveler in Rockford, Illinois, suggests going a step further and taking a practice trip before the actual vacation. In preparation for this trial, "give the family member who has special needs the opportunity to have control over something, like doing a planner that helps identify what to bring along, which things will be challenging, and how to handle them," she says.

"Along with reminding kids the adventure will be fun, keeping kids calm and focused means making sure you have the usual necessary items on hand—snacks, drinks, and entertainment," says Nicole Graham Wallace of Trendy Travels by Nicole—Dream Vacations in McDonough, Georgia. But parents can also use the experience to educate and emphasize appropriate behavior, like "explaining the role of the train personnel and the importance of staying in your seat or berth," suggests Sarah Salter of Travelmation dba Sarah Travels Happy, in Fayetteville, Georgia.

Amtrak's website includes information on wheelchair accessibility, traveling with service animals, and the discounts they offer for travelers with a disability as well as the companions who must accompany them. The requirements and restrictions are available at https://www.amtrak.com/deals -discounts/everyday-discounts/passengers-with-disabilities-discounts.html.

Savvy parents can take additional measures to make the ride even more autism friendly.

For example, on trips where such accommodations are offered—especially for trips longer than six hours—consider opting for a roomette or private room, advises Lisa Bertuccio of Vacation Your World in Otisville, New York:

This goes for neurotypical children under the age of ten and children with ASD. The reason: the train ride can be loud at times, with the horn blowing and the sound of the people walking and talking inside of the cars. Having a roomette or private room gives the parent a little bit more control over the environment yet allows the child to have more freedom. Also, these rooms have their own bathroom, which is a great bonus when traveling with kids in general. As with planes and buses, requesting to board first and exit first—to avoid waiting in line and dealing with crowds—is something that can be done when you first get to the train station.

That private room or roomette can be invaluable at mealtime, says Jennifer Hardy of Cruise Planners in Kent, Washington. "There is nothing worse than a child having a meltdown while you wait for a table in the dining car. With a private room, your meals can be delivered to you instead—no wait!—and an attendant will be there to assist you."

For rides where sleeper units are not offered or are sold out, Chandra Nims Brown of Cruise Planners—Global Travel Pros in Williamsburg, Virginia, suggests trying to secure seats in the rear of the quiet car or in a car with fewer passengers.

Whether it's a seat or a room, don't feel like you need to spend the entire journey in one place, say the experts. One of the advantages of this mode of travel is "your ability to move around. If your child is struggling to sit still, you can [get up and explore] other areas of the train," says Angela Romans of Rejuvia Travel in South Bend, Indiana. And Tara Woodbury of Escape into Travel in Leland, North Carolina, agrees: "Just like with any other type of transport, you will want to have plenty of distractions and if needed, tools for sensory regulation. A scenic car may be a great place for your child on the spectrum."

While train travel lets you admire the countryside—and is achievable for families with neurodiverse children, according to the experts quoted in this chapter—cruise travel offers equally memorable vistas and can be an easy entree into international travel. It's the subject of the next chapter.

PARENT AND CHILD PROFILE: REBECCA DA RE AND ROMEO

Rebecca Da Re is the mother of Romeo, a ten-year-old boy from London who has been diagnosed with autism and sensory processing disorder. She says that from age three, Romeo fell in love with Thomas the Tank Engine and his interest in trains grew from that point on.

"It wasn't ever just about the cartoon, he loved to watch the motion of the trains in the animation, learning the names and engine types of every train in the program," she says:

> He started collecting and playing with the physical toy engines. He would line them up and ask us to point to each one and he would test us on [both the fictional] name and the real-life train [model]. Then it

was Brio train sets, Trackmaster. . . . At age five or six, he was particularly interested in visiting specific stations and seeing particular trains and started videoing them. Romeo also took an interest in train maps and route networks; he would spend many hours researching maps to figure out which train routes he had not been on and train types/livery he hadn't seen.

Da Re says that now, at age ten, "Romeo is able to identify a train's model and stock [with] just a side glance, or a station by which trains he can see in the photo as he has memorized every route network. He inspires many other young people with his contagious excitement and passion for trains . . . and continues to astound most adults with his knowledge."

In addition, she says her son is a very active and "known" boy among the train networks in the United Kingdom, specifically Transport for London (TFL), a local government body responsible for most of the transport network in London, and that [when he was nine], he was one of their youngest volunteers at the Acton Depot. "We take trips across the country from Portsmouth up to Scotland. The National Railway Museum in York and the London Transport Museum are two of his favorite places to go. Recently, we visited the Glasgow Subway, which was on his list for a very long time. Whenever we are away on foreign holidays, visiting the local train lines is always on the itinerary."

Romeo also has a very active presence on social media ("RomeosworldofTFL") where he posts original content that documents his almost-daily train trips and discoveries. "On these journeys, we often meet other trainspotters and like-minded children who have become friends of Romeo, often through Transport Sparks, a Facebook group that was started by parents of autistic children that love trains. The group helps connect children and parents, and provides a safe, positive community and meet-up opportunities," she says.

In the United Kingdom, Da Re explains that neurodiverse children often travel wearing sunflower lanyards, which identifies them to station and airport staff. "Train drivers and station workers are almost always hugely accommodating to the children they see filming or jumping up and down with excitement when a train comes in." Unfortunately, she laments, other passengers are sometimes less accommodating, despite a campaign over the past two years across rail networks highlighting "invisible disabilities." "When trains are busy and crowded, it can be very overwhelming for an autistic child. You can send off for a badge that alerts other passengers that you or your child [should] be offered a seat . . . but sometimes having to convince a fellow passenger that a seat is needed without pointing out specifics remains an issue."

Unexpected delays and cancellations can also prove a huge area of disappointment and waiting around doesn't work well when you have autism, says Da Re. "As Romeo is getting older this part is getting easier and more manageable, but when he was younger, sadly there wasn't much we could do other than be prepared with his favorite foods and snacks. Pack a favorite toy, electronic device, or game—distraction, distraction, distraction."

More recently, the COVID-19 lockdown created a new kind of disappointment by putting a temporary halt to Romeo's railway visits. To occupy his time, "Romeo decided to draw the London tube map and using Google Maps, list every London hospital and mark each based on which tube station was closest. It actually ended up going viral and we received messages from around the world. In the end, it gave such a boost to rail staff that they put it up in stations, which kept Romeo upbeat and connected for that time."

Da Re believes Romeo's most memorable train rides were his first time traveling on a Pendolino train when the guard made a personal announcement to him [over the loudspeaker] mid-journey. Or more recently, on an Azuma train to the North of England which was his dream. "They made another personal announcement and let him see the driver's cab. That was his first time as a very young child riding on a DLR train, which is driverless so you can sit at the front, and it feels like you are driving the train. It was amazing."

Romeo's love of trains has brought him a sense of calm, she says:

> Trains tie into his neurological wiring. They have a great soothing motion that provides a huge amount of vestibular input, amazing for the sensory system. They are repetitive in motion, smells, and sounds. Predictable, routine . . . no surprises. Clear maps to follow which rarely change . . . doors opening and closing, wheels going round and round. The front of the train looks like a happy, smiling face. What we [neurotypicals] see as a simple practical solution [to getting around], they see way beyond.

But just as important as a sense of tranquility, trains have brought Romeo a sense of connectedness and hope for the future. Says Da Re:

From station dispatchers, train guards, drivers, railway signalers and engineers, I think overall what it means for Romeo is that he belongs, and wherever he goes, he knows that his thirst for railway knowledge and experience is not for nothing, that one day there will be a world of options available to him and he will be surrounded by such a like-minded community. He spent a long time feeling out of place in a neurotypical, confusing world. The railway is his world, his people. It's simple, safe, straightforward, relatable, and familiar. I'm not sure if any specific event or journey changed him in any way, but collectively it's cemented his desire to pursue his dreams of one day working in a career on the railways or for TFL.

CHAPTER 10 ACTION STEPS

1. Take advantage of a train's more liberal baggage weight restrictions to bring whatever familiar items your child might need.
2. Play up the excitement of train travel—especially if it ties into your child's passions. Remember that tram and subway travel might evoke similar excitement.
3. If possible, test for possible sensory overload in advance, as the terminal and the noise and motion of the train itself could cause problems for your child. A visit to the train station and possibly a short sample trip might be good indicators of what steps might be needed in advance of a longer holiday.
4. Have the child help in the planning of the rail journey.
5. As always, bring whatever's necessary (food, toys, etc.) to keep the child calm and engaged.
6. Consider a private room or roomette, which will provide distance from other passengers as well as easy bathroom access and an alternative to the main dining car. Alternatively, seek out seats in the rear of the train or a car with fewer passengers.
7. Take advantage of the ability to move throughout the train if kids get antsy. The scenic car can provide a variety of views and distraction.

11

SMOOTH SAILING
Cruise Travel

"Nate enjoys cruise ships. Our travel agent found a great cruise out of New Jersey to Florida with a couple of stops in the Caribbean. We live in Massachusetts, so we drive to New Jersey and hop on the cruise. Royal Caribbean is super helpful when we arrive at the terminal. I find a cruise rep with a yellow jacket, say 'Autism,' and away we go . . . smooth." – Michael Sokol, parent, Longmeadow, Massachusetts

Don't forget about the high seas when planning a vacation, whether your family's makeup is neurotypical or neurodiverse. For one thing, the logistics are easier. As Starr Wlodarski of Cruise Planners—Unique Family Adventures in Perrysburg, Ohio, tells her clients, "Cruises are great for families, especially for kids on the spectrum. You come back to the same room every night and eat familiar foods." In addition, children can move more freely aboard ship, which is a boon to those who might struggle to sit in a car or airline seat for an extended period, or who may want to escape an area teeming with overstimulation. Another bonus is flexibility: with cruising, you can choose from trips of various lengths and price tags, ships of different sizes, and a large selection of departure ports.

Still, there are challenges to consider. For example, "traveling on the water can bring additional quandaries to be addressed such as motion sickness, confined spaces [like cabins], as well as protection from deck railings. It would not be advisable to have a balcony stateroom without taking extra precautions to ensure the child cannot access the balcony alone," says Betty J. Burger, recently retired from BGB Travel and Leisure, LLC, St. Marys, Georgia. Adds Carl W. Johnson of Adventures in Golf (incorporated as Hamill Travel Services, Inc.) in Amherst, New Hampshire: "Although different than other methods of travel, the same concerns exist. Crowds getting on and off cruise ships, onboard lighting, various noises, food lines, crowded pools. While this may not be an issue for some, it could be a real problem for others with heightened sensitivities."

Travel advisors were divided as to the size of the cruise you should choose if you have anxious, inflexible, or neurodiverse children. Says Holly Roberts of Frosch Travel—The Travel Center in Hanover, Massachusetts: "I would not recommend a large cruise ship since there could be several thousand people aboard, lots of noise, busy swimming pools, and large restaurants." And Tara Woodbury of Escape into Travel in Leland, North Carolina, warns, "If your child has demonstrated unsafe behaviors such as running off, a cruise ship is not necessarily the best way to vacation. [If you must cruise,] a smaller ship may be better suited for your family if eloping is an issue."

But other agents disagree: "A larger ship may seem like it would be overwhelming, but it may be better for your child with autism. There will be more passengers, but they will be spread out. There will be more activities planned and the crowd you are surrounded by can essentially be smaller," says Marian Speid of Dream Vacations, Fair Lawn, New Jersey. Adds Jennifer Hardy of Cruise Planners in Kent, Washington: "The larger the ship, the more flexibility you will find. This is perfect for avoiding specific sensory stimuli or seeking out the just-right foods your child will eat."

Most agree that you should dip a toe in before jumping in with your whole body. In other words, don't assume that if you want to try cruising, you have to be away a full week or longer. Some cruise lines offer two-, three-, and four-night cruises, and if that's too long, consider a three-hour sightseeing cruise on a local lake or river, or even a ferry ride. "Start with a shorter cruise and work your way up to lengthier itineraries farther from home as your family gains comfort on board," says Hardy. Sarah Marshall of TravelAble in Midlothian, Virginia, sees it as a launching pad: "Cruises can be a great first step in traveling to other countries. Several cruise lines have autism programs that allow families to include their children in the kids' club, while offering aids and tools to help children if they need to stay in the cabin."

Big ship or small, short cruise or longer, domestic or foreign ports of call—these decisions come down to your instinct, your child(ren), and some experimentation. Luckily, many Certified Autism Travel Professionals (CATPs) and cruise lines are attuned to the needs of families with neurodiverse children.

"Let your travel advisor know in advance that you have special needs that must be accommodated," says Hardy:

This not only helps them to match you to the right cruise line and itinerary, but it also allows them to alert the cruise line in advance to prepare for your arrival. Most cruise lines can cater to dietary restrictions, but only a few have the necessary training for working with individuals on the autism spectrum. Knowing your needs in advance allows everyone involved to truly assist in making your vacation a successful experience.

Consider these other tips if you are planning a holiday at sea:

PREPARING THE CHILD FOR THE CRUISE

Taylor Covington of *The Zebra* advises using Social Stories (see chapter 6), videos, and similar tools to prepare your child for this new mode of transportation. "Plus, ships can be loud. If your child has auditory sensitivities, be prepared with noise-canceling headphones and other modifications beforehand [to] prevent [possible] meltdowns," she says.[1]

Some cruise lines allow for a trial run to prepare your child, according to Brenda Revere of C2C Escapes Travel in Bedford, Indiana:

The cruise lines that offer autism programs on board include Royal Caribbean, Celebrity, Disney, Carnival, Princess, and Norwegian. Before COVID-19 hit, Royal Caribbean would allow [you to schedule an] appointment to board the liner when docked so families could see the ship before sailing. You could tour a stateroom, see where the kids' club was located, visit the pools and entertainment areas. Now, you can prepare by watching YouTube videos of the ship you will be sailing on and if you have any questions, you can call Guest Services and ask for the Special Needs department.

CHOOSING AND BOOKING YOUR CRUISE

When you reserve a cabin aboard a ship with crews trained to work with and accommodate children on the spectrum, it gives parents a much-needed break and some alone time during the day and early evening so they can enjoy a vacation too, says Jonathan Haraty of Adventure Luxury Travel in Whitman, Massachusetts. "[But] yachts are another option if you don't get seasick on smaller [vessels]. You can rent the whole boat for yourselves with just a captain and chef or with a full crew. It is a great way to see the Caribbean or the Greek Islands."

Another option is renting a houseboat, which Haraty says is a wonderful choice for a private vacation on a lake, but most do not come with a captain or crew. "If you have a sailor in the family or want to learn how to pilot a boat, the houseboat is probably one of the best options to try out something different with your child on the spectrum. It's just your family onboard and you have full control over your child's environment," he says. (There is a section on houseboats later in the chapter.)

If you do choose a regular ocean liner, Angela Romans of Rejuvia Travel in South Bend, Indiana, recommends traveling during the off-season to save some money and lessen the size of potential crowds. And Nicole Graham Wallace of Trendy Travels by Nicole—Dream Vacations in McDonough, Georgia, advocates selecting an itinerary and cruise ship that is "family friendly . . . be sure that the kids' program has trained professionals that can properly care for children on the spectrum."

"To that end, booking an Autism on the Seas cruise [profiled later in this chapter] is a great option as they provide staff-assisted cruises," says Olivia McKerley of Elite Events & Tickets in Grovetown, Georgia. "Also, many cruise lines like Royal Caribbean, Carnival, Celebrity, Disney Cruise Line, and Norwegian have specially trained staff and offer activities specifically for families with autistic children. They understand things like having the right lighting, limiting crowd size, providing sensory friendly movies and toys, priority check-in, and more."

If you do not use a CATP to book your cruise, let the cruise line know prior to booking that you will be traveling with a child with autism spectrum disorder (ASD) and notify them that you would like express boarding. Also advise them of any special food requests or restrictions, warns Lisa Bertuccio of Vacation Your World in Otisville, New York. Along those lines, be sure to request a phone [or pager] so the dining and kids' club staff can communicate with you throughout the cruise, says Marian Speid. "This request should be made ahead of time, and you may want to keep your confirmation on hand during embarkation."

CABIN CHOICE AND LOCATION

As with airline seats, the choice of cabin location is important on an ocean liner, especially considering the presence of noise and ship movement. Lisa Bertuccio recommends the aft of a ship, which tends to be quieter than the front or midship. As for motion sickness, "some feel that the aft—although quiet—can have a bit more movement, so if that poses an issue, opting for a cabin in-between the midship and aft [sections] of the ship would be best," she says. Angela Romans adds: "If your child is sensitive to loud noises try and avoid locations near the theater, engine room, and dining rooms, or pack noise-cancelling items."

For those whose children need more space, the average cabin size (160 square feet) might feel a bit claustrophobic. In such cases, Marian Speid recommends reserving a bigger cabin or two adjoining cabins.

WHAT TO PACK

As with all modes of travel, a go-to bag is a must. Speid recommends the usual items—chewing gum, weighted blanket, pillow, headphones with music or games, weighted Lap Pad, fidget toys like Silly Putty or a worry stone, bottled water, a favorite toy or stuffed animal, healthy snacks, earplugs, and clothing for layering—along with some specifically for the warmer weather, such as a hat and sunglasses, sunblock, and Sea-bands and Dramamine for possible seasickness. (Read more about motion sickness remedies in chapter 8.)

And as always, come prepared with safety items. Authors Jesemine Jones, a school social worker, and Ida Keiper, a retired special education teacher, recommend purchasing a portable battery-operated motion detector alarm system to use on the doors in your cabin. "This is especially important if you are in a cabin with a balcony. This system will sound if a door is opened," they say.[2]

DINING

There's usually some variety in dining aboard ships, especially on larger, more modern vessels. This means you can choose from buffets and informal settings as well as more traditional dining rooms. Some families bring familiar foods from home that don't require heating as cabins don't have cooking facilities for safety reasons. However, the selection of dining choices aboard should satisfy most parents with neurodiverse children.

Request your dinner table to be located near the entrance of the restaurant, suggests Jennifer Hardy:

> The corners tend to be a little quieter, and the proximity to the entrance will allow you to make a quick getaway if needed. Some cruise lines will offer a family dining experience—serving the children faster and whisking them off to the youth activities center, while the adults can enjoy a more leisurely meal. You can also request a private table just for your family, so you do not have to share with strangers.

"Cruise lines will also accommodate those with a limited palate and will allow in-room dining, if that is a concern," adds Belinda S. Stull of Travel Leaders/All About Travel, LLC, in Hagerstown, Maryland. Or you could request permission to take food from the dining area in case of emergency, says Angela Romans.

EXCURSIONS

When in port, it might be advisable for families on the spectrum to limit the number of excursions they take and research which ones would be better suited to their child. "Some cruise lines have their own private islands where you can just kick back at the beach and/or waterpark," says Tara Woodbury. "This may be a better idea than say, a shopping tour. You always have the option of staying on board the ship during days in port. This usually means the pool and other facilities [will be] less crowded, so it may be a great time for your family to enjoy some of those amenities," she adds.

OTHER CONSIDERATIONS

Making your child feel comfortable aboard ship is a must. Jesemine Jones and Ida Keiper recommend requesting an orientation tour early in the cruise in order to familiarize your child with the liner's public spaces.[3]

Lifeboat/muster drills can seem chaotic and the crowds traumatic to those with special needs. Experts' advice includes making sure your child has a properly fitting life jacket and having your travel advisor call in advance to make special arrangements, such as asking if there's an alternative to attending. If not, then "explain the muster drill. Bring a friend to the muster drill. And spend extra time explaining water safety," says Sarah Salter of Travelmation dba Sarah Travels Happy in Fayetteville, Georgia.

One thing that travelers love about cruising is that this mode of transport also provides the accommodations. However, for families on the spectrum traveling by air or road who need to find a place to spend the night, the next chapter covers how to handle stays at hotels and vacation rentals.

MODIFICATIONS FOR CRUISE PASSENGERS WITH AUTISM

Royal Caribbean International has long been considered one of the pioneers in making cruising accessible for all. Here is their list of modifications for neurodiverse travelers, taken with permission directly from the Royal Caribbean website.[4]

Note: Programs can change without notice; check directly for current modifications. Also, check with other cruise lines about similar programs—their contact information can be found later in this chapter—and, if you need to bring your service animal aboard, check directly with the cruise line in question regarding their policies and procedures.

Autism-Friendly Ships

Royal Caribbean offers an autism-friendly initiative for families living with autism, Down syndrome, and other developmental disabilities. This includes sensory friendly films and toys, dietary menu options, and overall autism-friendly training for Adventure Ocean staff and more.

Royal Caribbean offers a wide range of autism-friendly products and services, and they include:

- Priority check-in, boarding, and departure
- Special dietary accommodations including gluten-free and dairy-free

- Adventure Ocean flexible grouping by ability for children three to eleven years old
- Adventure Ocean toilet-trained policy exception
- Pagers/phones for parents of children in Adventure Ocean program while signed into our care (subject to availability)

Availability: These autism-friendly products and services are available on all cruises.

In addition, Autism on the Seas (AotS) offers "Staffed Cruises" catering to families with children, teens, and adults with autism and other developmental disabilities. These "Staffed Cruises" include:

- Extra staff professionally trained in caring for individuals with developmental disabilities (provided at one staff member for every two or three special needs guests)
- Assistance with Royal Caribbean's products and services
- Specialized respite sessions
- Private activities and sessions

Availability: AotS "Staffed Cruises" are offered on select cruises.

The Autism Channel

Royal Caribbean International is the first in the hospitality industry to offer complimentary on-demand access to exclusive content from The Autism Channel® on board most of its ships.

The Autism Channel is a streaming television service providing information and resources to families and professionals supporting and improving the lives of people with ASD. Educational and entertainment programming on The Autism Channel ranges from a look inside the daily lives of families with children on the spectrum, to interviews with medical and legal professionals.

The Autism Channel is available on all Quantum, Oasis, Freedom, Voyager, and Radiance class ships. To access The Autism Channel content, select The Autism Channel folder under OnDemand Movies from the interactive television menu.

Autism-Friendly Films

Autism-friendly films are presented in a low-lit and low volume environment. Guests are encouraged to freely talk and walk around during the film. These films will be offered on all sailings on Oasis class ships, and on other ships when there are autism groups or at least five individuals with autism on board. Dates and times will be highlighted in the Cruise Compass.

Autism-Friendly Toy Lending Program

Autism-friendly toy lending is available in our Adventure Ocean Youth Program on all ships. Upon request, we will provide parents with a tote bag of autism-friendly toys that may be used in Adventure Ocean or in their stateroom. Parents may select other autism-friendly toys that are more appropriate for their child. Examples of autism-friendly toys include non-toxic crayons, markers, watercolors, building blocks, dominoes, and picture books.

Autism-Friendly Activities

We offer activities for children of all abilities. Families are encouraged to consult with Adventure Ocean staff regarding any special needs that their children may have in order to identify which activities are appropriate for their child and any possible modifications.

Cruising Social Story

A Social Story is a written or visual guide describing various social interactions, situations, behaviors, skills, or concepts. These Social Stories help individuals with autism to better cope with social situations. Royal Caribbean offers a Cruising Social Story about cruising to help families with autism prepare for their vacation.

LINKS TO OTHER CRUISE LINES' SPECIAL NEEDS PROGRAMS

For more information about the special needs programs on the other lines cited by CATPs as leaders in ASD-friendly cruising, here are their links:

- Carnival Cruise Lines: https://www.carnival.com/en-US/about -carnival/special-needs/children-special-needs
- Celebrity Cruise Lines: https://www.celebritycruises.com/int /special-needs/guests-with-autism
- Disney Cruise Lines: https://origin-dscribe.s3.amazonaws.com /disney-cruise-line/guest-services/guests-with-disabilities/DCL -Information-for-ASD-13-Dec-2019.pdf
- Norwegian Cruise Lines: https://www.ncl.com/uk/en/about /accessible-cruising
- Royal Caribbean Cruise Lines: https://www.royalcaribbean.com /experience/accessible-cruising/autism-friendly-ships

SUPPLIER SPOTLIGHT—AUTISM ON THE SEAS

If it feels daunting to piece together the various cruise strategies offered by the CATPs on the previous pages, AotS is a one-stop shop that can make all those arrangements on your behalf and even hold your hand during the cruise. According to their website at Autismontheseas.com, this international organization has been in collaboration with Royal Caribbean International since 2007, developing cruise vacation services to accommodate adults and families living with children with special needs including, but not limited to, ASD, Down syndrome, Tourette syndrome, cerebral palsy, and all cognitive, intellectual, and developmental disabilities. These services quickly expanded to other cruise lines.[5]

"Our services and professional volunteers provide families the additional support and aid that transforms a trip into a vacation," says Michael Sobbell, AotS chief executive officer and founder.

Staff-assisted cruises are selected from regular cruises throughout the year, and AotS says they "assist adults and families by accommodating the typical cruise services, as well as providing specialized respite and private activities/sessions that allow guests the use of the ships' entertainment venues in an accommodated and assisted manner." Their staff, whom they advertise as experienced, background-checked, and sanctioned by the cruise lines, accompanies neurodiverse guests on their cruises to provide "amazing" vacation and travel experiences on board Royal Caribbean, Celebrity, Norwegian, Disney, and Carnival Cruise Lines.

These staff-assisted cruises operate almost every week throughout the year and cater to special needs guests regardless of age: children traveling with family, teenagers traveling with family, and adults traveling with family or a group home. The staff also coordinates private activities and respite services, which range from one to three hours at a time. Private sessions are assisted by staff at most of the ship's venues. Seats are reserved at shows, so guests do not need to arrive early. Guests also receive a private muster drill, expedited boarding and disembarkation, reserved seating for all meals, and more. AotS also offers such perks to extended family and friends, and neurotypical siblings and accompanying friends can attend respite sessions.

On the AotS Testimonials Facebook page, parents were effusive about their experiences, commending the company for allowing families with special needs children to actually have a vacation by removing stress, being attentive, and facilitating friendships between like-minded travelers. Many emphasized how eager they were to book again.[6]

For guests who prefer to cruise on their own, AotS provides a Cruise Assistance Package, a service that is offered free of charge just by booking through the company. This option allows guests to pick the cruise of their choice from various cruise lines, and AotS will provide "guidance and consultation on any concerns, needs, or issues guests may have as well as accommodated boarding, disembarkation, and muster drills, providing information on guests' children for the cruise line youth staff, excursion reviews from past guests, special name tags for children, AotS T-shirts, and more." Guests pay the same rates found on the cruise lines' websites.

For those who can't afford to cruise, the AotS Foundation offers two types of financial assistance. For families that have a child diagnosed with a disability and wish to vacation on AotS Staff-Assisted Cruises, the Vacation Assistance Program awards varying amounts to immediate family members only—parents/guardians, qualifying children/adults, and their siblings. The foundation evaluates applications individually to determine the amount of the financial assistance to be awarded and based on funding available at time of awarding. Families are awarded in the order in which they register. Vacation Grants, offered through the "Sponsor a Family" program, allow individuals, families, and businesses to create and donate to a Vacation Grant, which will be awarded to families to receive AotS services. The families are chosen based on the demographic options that are selected by each donor, and again, the order in which families register.[7]

You can register for their financial assistance program on their website at www.autismontheseas.com/financial-assistance.

TRAVELING ON A HOUSEBOAT

For families who prefer to sail on a smaller scale, or just enjoy being near the water, a houseboat rental might be the perfect compromise between cruising and a hotel stay.

Explains Theresa Jorgensen on the SixSuitcaseTravel website, houseboats are moored on lakes around the world and come equipped with a full kitchen—often including a grill—as well as bathroom, living room, and sleeping areas. The upper deck provides great sunbathing space. Some even have swim slides to get into the water faster.[8]

It's up to you to provide the meals, which is one way to ensure you'll be serving familiar food your child will eat. You also provide the schedule and entertainment, which means it is easier to stay on routine, even if it's less of a vacation for parents who are still on twenty-four/seven call. But since houseboats can sleep between six and thirty passengers, the additional bedrooms can provide space for any staff you might choose to recruit, such as someone to cook and/or watch the kids so you can enjoy some time alone. Alternatively, you can invite along family and friends to share the experience (and the work), including keeping the kids safe. Emphasizing water safety is a huge issue when traveling with children on water, whether they be neurotypical or neurodiverse.

Some houseboats remain moored while others are motorized (and like a rental car, are gassed up before you depart), which allows you to change the scenery when desired, whether it's to find an inviting beach or a prime fishing spot. You can even rent water toys like a tube, kayak, or paddleboard to increase your fun. Marina staff are happy to teach you how to operate and navigate the houseboat when you pick up your rental.

There are some security measures to consider when renting a houseboat. On their website, Cottonwood Cove Marina suggests bringing prefitted life jackets from home (even for infants!) and getting used to wearing them before leaving home; understanding the difference between a floatation device (helps the wearer stay afloat) versus a life jacket (designed to keep child from drowning); and learning about the vessel's safety features from the marina staff, which may include railing guards that keep passengers from falling overboard, sliding doors with automatic lock features, and gates with automatic latches on the boat's exterior that keep unsupervised children from taking an unprotected swim.[9]

In her article "17 Tips You Must Know Before Renting a House-boat," fitness and health guru Kiana Tom lists safety measures she's learned from personal experience, such as videotaping the marina owner's explanation of how to operate the boat, how to anchor the vessel correctly, what to know about the lake you plan to visit, and much more. It's worth a read at https://kiana.com/17-tips-must-know-before-renting-a-houseboat/.[10]

Houseboating.org also reminds renters to "only swim about the houseboat when the engine is off and the houseboat is stationary, to en-sure the propellers aren't in motion." In addition, they warn not to tube/wakeboard/ski behind the houseboat. "The houseboat's generator is also something to be aware of in regard to safety concerns. As the generator emits carbon monoxide, you'll want to be sure it's off when swimming behind the houseboat or using the slide, as well as refraining from running it all night in order to steer clear from carbon monoxide poisoning."[11]

Rental prices vary depending on the size of the houseboat and the des-tination chosen, and of course, the amount of fuel you'll need if you plan to sail. According to Jorgensen, some lakes require you to bring or rent an additional motorized boat to tag alongside your houseboat rental for safety reasons, but typically the associated costs are reasonable. She says the other charges that may not be included in your rental cost include propane, park entrance admission, environmental fees, etc. Ask your agent what other sums you can expect to pay (if any) at the location you plan to visit.[12]

You can find listings of houseboats for rent on many sites across the internet including www.houseboating.org and www.airbnb.com.

CHAPTER 11 ACTION STEPS

1. Use Social Stories, videos, picture books, etc., to prepare for the cruise and to emphasize water safety.
2. Know your child. Will crowds at embarkation and disembarkation be a problem? How about onboard lighting, food lines, crowded pools? Select the size of the ship and length of cruise accordingly.
3. See if your cruise line permits trial runs or advance tours of the ship. Otherwise, YouTube might suffice.

4. If unsure, start with a short, two- or three-day cruise first, or even a three-hour sightseeing cruise on a local lake or river or a ferry ride.

5. Make sure your cruise is kid friendly. Seek out ships with programs for families on the spectrum, or companies that specialize in group ASD cruises like AotS.

6. Cruising alternatives to consider include private yachts and houseboats.

7. Advise your CATP or the cruise line (if you're booking directly) about your child's diagnosis and any dietary and other special needs.

8. Choose cabin locations that minimize noise and motion. Consider larger or two adjoining cabins if your child will need the space.

9. Avoid cabins with balconies if you cannot provide precautions to ensure the child can't access the balcony alone. There are alarms you can purchase to place on interior doors of the cabin.

10. Have a plan in place for the lifeboat/muster drill.

11. Structure dining choices around your child's needs, whether that means visiting the informal dining rooms, ordering in-room dining, or bringing back food from the dining room to eat in the cabin. If you do select the dining room, choose quieter areas with good proximity to the exit.

12. As with all modes of travel, bring noise-canceling headphones if the child has auditory sensitivities, as well as whatever's necessary in case of motion sickness.

13. When in port, select excursions in tune with your child's needs or stay onboard and enjoy the less crowded facilities.

Part IV

ONCE YOU'VE ARRIVED

12

STAYING THE COURSE

Various Accommodations Options

"Sometimes having an environment that you can control is easier than staying at a hotel. Consider a vacation rental through Airbnb. Renting a house provides you with a kitchen to cook in but you can still eat out when you want. You'll also have the number of bedrooms you need. Most owners are happy to accommodate you if you need a fan or a highchair." – Michelle Zeihr, parent, Barrie, Ontario, Canada

Once you've arrived at your destination—either by air, rail, car, or ship—you'll need to have somewhere to stay, somewhere that is understanding of, and can cater to, your child's particular needs. It takes a great deal of preparation and research to find the right accommodations. What might feel luxurious to parents may be interpreted by the anxious, inflexible, or neurodiverse child as a chaotic mélange of unfamiliar and meltdown-triggering sights, sounds, and smells that ultimately ward off comfort and sleep.

"Ideally, you will want to stay at a hotel that your Certified Autism Travel Professional (CATP) has personally visited," says Jennifer Hardy of Cruise Planners in Kent, Washington:

They will be your "go-to" when it comes to learning the important details, like whether there is visual or audio stimuli that could impact your experience. These are things that the internet cannot tell you. Your travel professional can also recommend accommodations with the best pools, lowest crowds, full kitchens, and more, [to help] families' . . . expectations at the time of booking . . . match their actual experience.

Clearly, booking with a CATP can be invaluable, even though their opinions may differ on whether a full-service resort or a vacation rental would be the better choice for those on the spectrum. However, if you choose to book accommodations on your own, what follows are some tips to help smooth the way. Please note that while this chapter covers hotels, resorts, and vacation rentals, other accommodations options include dude ranches, houseboats (see chapter 11), and camping (see chapter 13).

CERTIFICATION STANDARDS

In chapter 5, I discussed the difference between autism-friendly and autism-certified venues, labels to be aware of when seeking out accommodations. Now might be a great time to discuss the origins of the Certified Autism Center (CAC) and Advanced Certified Autism Center (ACAC) designations, as well as the CATP program.

The International Board of Credentialing and Continuing Education Standards (IBCCES) is the organization behind these designations. Meredith Tekin, BCCS, the organization's president, says IBCCES identified the need for trained and certified travel and recreation options for individuals with autism and their families/friends

> as a natural extension of the work we have done for twenty years in education, healthcare and corporate settings. We consistently were approached by adults with autism or parents of ASD children, as well as destinations and attractions looking for solutions. IBCCES offers training and certification to hotels, resorts, and attractions, but we also saw a need for travel agents to have more knowledge and expertise in this area. Thus, the CATP [designation] was created to help travel professionals better understand and feel comfortable planning trips and assisting individuals and families with these needs.

To increase awareness of the program, both to travel advisors and to the public, Tekin says that IBCCES has a public registry on ibcces.org listing all their certified professionals and organizations, so anyone can identify and verify certified resources:

> We also promote our travel and recreation partners on AutismTravel .com, including CATPs, as another resource for parents and individuals to locate trained and certified professionals and destinations. We have new CATPs register every day, but our focus is broader in terms of en-

suring there are trained and certified hotels and attractions for individuals and families to visit all over the world. CATPs are an important part of that goal but we also have a large focus on the locations themselves to ensure they can welcome guests with all abilities once they arrive.

The organization also created the IBCCES Accessibility Card (IAC) program to help certified destinations and attractions better manage their accessibility programs and to help individuals communicate their needs to those locations, says Tekin. The Accessibility Card is not specific to one need or disorder; anyone with a need or accommodation request can use the IAC when visiting participating locations. The Accessibility Card doesn't guarantee specific accommodations at a given location but provides a way to communicate those needs (not specific to a diagnosis but specific to the needs the location may be able to meet or accommodate) as well as streamline the process onsite. More information is on their website at https://accessibilitycard.org as well as in chapter 14.

As for the future of IBCCES, Tekin says that the organization will continue to work with hospitality and recreation partners in the United States and across the globe, as well as corporate neurodiversity partners, hospitals, therapy clinics, law enforcement, and education settings:

> COVID-19 of course changed things for a lot of travel-related organizations in 2020 and 2021, but we continue our mission to provide a verified, comprehensive certification process and we continue to receive inquiries daily about how to implement these programs in various settings. We have worked with our existing partners to adjust and continuously improve programs during [the pandemic] to ensure they service guests to the best of their ability as they reopen or expand operations. As a credentialing board, we consistently update our training content and share best practices with partners.

SETTING THE STAGE

As with all aspects of travel, before choosing accommodations, preparing the child for a change in sleeping arrangements is key. Olivia McKerley of Elite Events & Tickets in Grovetown, Georgia, suggests Social Stories (chapter 6) "as a great way to help kids understand what to expect and how it will be different from being at home, but that it will all be okay." Adds

Nicole Thibault of Magical Storybook Travels in Fairport, New York: "You may want to go on the hotel's website and show your child pictures of the lobby, room, even the outside of the building. With those pictures, you can [discuss] the place you'll be staying and maybe even talk about how long you'll be there. This can give your child a sense of calm and ease about the trip."[1]

"Understand and prepare [the child in advance] for the effects of a disrupted routine," says Taylor Covington of *The Zebra*. "Your child may have trouble sleeping in the hotel. Bring all the aids you would normally use at home but be willing to be flexible. Extra naps, earlier bedtimes on subsequent nights, and an attitude of flexibility will be helpful."[2]

HOTEL (LARGE OR SMALL) VERSUS APARTMENT/HOUSE RENTAL

There's a definite dichotomy of opinion regarding the most suitable accommodations for children on the spectrum. Both sides have valid points but, ultimately, parents know their child best. Whatever your choice, as always, it is important to let the hotel or resort know your circumstances up front.

Carl W. Johnson of Adventures in Golf (incorporated as Hamill Travel Services, Inc.) in Amherst, New Hampshire, feels that rental houses or apartments are often better suited for people on the spectrum than hotels or inns:

> This is simply because most hotel rooms are small and don't offer a "getaway" place for someone who needs a quiet area to settle down after a long day of travel. Many suites offer multiple rooms, but not all travelers can afford the price of a larger hotel suite. A rental or Airbnb is always preferable when possible. Even a villa or condo on a hotel resort property is far better than a single hotel room since there will be more space, and likely a private area for people to remove themselves to "recalibrate."

Airbnb and rental apartments offer some advantages over hotels in that there are no noisy hotel hallways or neighbors in the next room, says Jonathan Haraty of Adventure Luxury Travel in Whitman, Massachusetts. "They are also more private because you'll usually be the only ones in the unit," he says.

Having only family members around is a clear advantage. However, Haraty points out that one drawback could be location; house/apartment shares may not be as centrally located as a hotel. "Also, there have been issues with guests not receiving the quality or amenities advertised. Airbnb is working on this issue but beware, just in case," he warns.

On the other side of the debate is Georgiann Jaworskyj of Custom Travel Services, Inc. in Merchantville, New Jersey, who says a full-service resort would always be her top pick, especially Beaches Resorts or Disney World, where the staff has been trained to assist. Jennifer Hardy of Cruise Planners in Kent, Washington, agrees: "I book well-known, larger chain hotels and lean away from Airbnbs or lower-star accommodations. This helps reduce the variability of the individual room my clients will stay in. It also sets a clear standard for service at the location," she explains.

"When booking a hotel room, definitely look for a hotel that is autism friendly," says Tara Woodbury of Escape into Travel in Leland, North Carolina:

> They can be hard to find but some will be CARD-certified, which means the staff receives annual training on providing a better experience to guests and families with special needs. Some hotels offer "safe kits" for the room including things like door alarms, corner cushions, etc. If you need something like safety rails for the bed, ask if they can be provided. If not, there are rental companies who can provide such items. CARD-certified hotels may also offer downloadable social books to help with your stay. Some, like the Tradewinds Resort in St. Petersburg, FL, offer kids' clubs with autism- and special needs–friendly activities. Hotels that offer room service or to-go meals might be a good fit for days when your child has had enough time around other people.

And then there's a middle ground. "We stay either at a hotel or a condo with room service when available," says parent Adrienne Query-Fiss of Seattle, Washington. "Sometimes our son just can't handle another noisy outing and needs a quiet meal in the room with his special toys and a movie."

DOING YOUR RESEARCH

As always, doing your homework reaps rewards in a trip enjoyed by all. "You want to make sure you are booking a hotel or apartment in an area that meets your needs—whether that means there's a doctor nearby, you're on a quiet street, or the accommodations are close to public transport," says Sarah Marshall of TravelAble in Midlothian, Virginia. Luckily, this information is as close as your internet connection. Advises Sandra Chu of The Enchanted Traveler in Rockford, Illinois, "Look at different views of the property. Try Google Maps or something similar to see the property's location in relation to its surroundings."

"I always make sure there's an outside space. I look for swings, pools, or Jacuzzis as amenities," says Brooklyn parent Delight Dolcine.

WHAT TO REQUEST WHEN BOOKING

Size and location of accommodations are additional issues to consider and request in advance. So are amenities. "Is a balcony acceptable? Does the hotel have a pool? What about the restaurants—are they able to accommodate children?" asks Belinda S. Stull of Travel Leaders/All About Travel, LLC, in Hagerstown, Maryland.

Again, there is some debate about location: Should travelers with children on the spectrum ask for rooms close to or far from the elevators? Likewise, should you book a main level room or something on an upper floor? There are the pros and cons to both but, as always, consider your particular child's needs and predilections when booking.

"Choose a second-floor room away from the stairs if your child is a wanderer," says Covington. "This will give you the ability to find your child quickly if they do manage to get out of the door without your knowledge and will prevent injuries." Conversely, Holly Roberts of Frosch Travel—The Travel Center in Hanover, Massachusetts, suggests a small hotel with a room on the first floor "where you would have easy access to the outside for walks." New York–based clinical psychologist Dr. Ellen Littman strikes a happy medium, suggesting that parents request a room on a lower floor near the elevator in a hotel with a coffee shop and pool, but no casino. "Children shouldn't have to start the day frustrated by how far they have to walk to the elevator, how long they have to wait for an elevator, and even how long the elevator ride takes. The goal is to minimize children's discomforts and parents' levels of frustration," she says.

Whichever location you choose, "getting adjoining rooms gives you the ability to set up one room with your child's special needs in mind while having another room where you can stay up watching TV or otherwise function without disrupting your child's sleep," says Taylor Covington. "Make sure, however, that the room is equipped with locks that your child cannot access, so you are not left to deal with a child who has locked you out of his or her sleeping quarters. . . . From locks on the sliding doors that access the patio to alarms for kids who wander, consider a hotel that has room security as part of its amenities," she says.[3]

While Taylor Covington suggests in-room security, Lisa Bertuccio of Vacation Your World in Otisville, New York, urges parents to seek out hotels and rental properties offering alarm systems, pool alarms, gates, and a video security system throughout the property in place in case your child takes off.

Access to kitchen facilities is also something to consider when choosing your accommodations. "If your child is a picky eater, a kitchenette can be a lifesaver on your trip, so look for a hotel room that has one," says Covington. In addition, "consider a hotel with a continental breakfast that offers more than just muffins and cereal, so your child starts off the day with a healthy breakfast," she suggests.[4] If a kitchenette isn't available, then opt for a hotel with rooms equipped at a minimum with a mini fridge and microwave, adds Bertuccio.

The topic of kitchen access is vital to Michelle Zeihr, a parent from Barrie, Ontario, Canada:

> Thessaly was an extremely picky eater. She only ate a handful of things and sometimes, it had to be a specific brand. If we were going to a resort, we would bring whatever food we could from home. We would always try to have a fridge in the room, and I would contact the property in advance to see if they offered milk cartons that we could keep handy. Then I would google what kinds of stores were around [that sold] acceptable foods for her, like strawberries, crackers, mac and cheese, and Pediasure, all of which I could keep in the room whether we had access to a fridge or not. I'd also research what restaurants were close by, like McDonald's and Domino's, in case the hotel couldn't provide food she'd eat. We would bring along some cans of Campbell's mushroom soup, the ready-to-serve variety—which she loved—and some small bowls with lids and a can opener. We'd open the soup in the room, bring it to the restaurant, and ask them to heat it up for her. Most places were happy to oblige.

Meanwhile, some kids on the spectrum relish hotel dining. "Nate truly loves the Hilton Hampton Inn line of hotels," says his father, Michael Sokol from Longmeadow, Masaschusetts. "They let you make your own waffles. Waffle-making is huge for us because Nate eats waffles for breakfast religiously! He also has a love for hotels with pools. I have even asked to access the pool earlier than its scheduled opening and we have never been denied."

A few other tips from Zeihr:

> When booking your accommodations, you should find out whether the bathroom will have a tub or shower. If there is only a shower available, we bring a small blowup pool from home and stick it in the bottom of the shower for the girls to bathe in. They love it. Another concern is [hallway] noise. My children have fans in their rooms at home for white noise, so we try to find accommodations where the air conditioning is loud or there's a bathroom fan that we can turn on to drown out noise.

And don't forget distractions, says Angela Romans of Rejuvia Travel in South Bend, Indiana. "Ordering Wi-Fi for the room is always a good idea as [most] children tend to enjoy playing games, listening to music, and playing on devices which connect to the internet."

ONCE YOU ARRIVE

If your child is sensitive to crowds, checking in might be a traumatic experience. Jonathan Haraty warns that

> hotels are busy, and their lobbies can [be noisy], so . . . if you have a car with another adult traveling with you, you may want to leave the child in the car with him or her and check in alone, and then have your child . . . taken directly to the room. Many hotels now have the ability for you to check in online or on your cell phone and even provide a digital room key on your cell. By using these tips, you can eliminate visiting the lobby and waiting your turn in line with your child.

Dr. Tony Attwood, a professor and one of the world's leading autism spectrum disorder (ASD) authorities, emphasizes that sleeping arrangements will be critical for the neurodiverse child:

It will be important that the person with ASD has a retreat within the accommodation, either their own bedroom or somewhere they can go to avoid contact with other people. This can be not only to relax but also to avoid prolonged social interactions that can be exhausting for him or her. If they have trouble falling asleep [and a second bedroom isn't available], moving their bedding into a closet might be another option.

Attwood also advises that the child may need extra help during time outside of the room:

It is also important to recognize that those with ASD have difficulty [interacting]. Just saying to the child to go and make friends is not easy [for them to do]. There may have to be structured or organized activities that [he or she] can join in. Should there be any recreational pursuits such as sailing or swimming, then whoever is providing the guidance to that group of children would need to know about the characteristics of ASD as expressed in that particular child.

Finally, "if food access, allergies, or pickiness are concerns, consider requesting special cleaning and grocery delivery," says Sarah Salter of Travelmation dba Sarah Travels Happy in Fayetteville, Georgia.

WHAT TO BRING ALONG

Keeping the hotel room or vacation rental comfortable for the child with ASD often means bringing familiar items from home, items that are easy to use, and whatever ensures safety while away from home.

"Bedtime can be a challenge for anxious children even at home—add a strange room and an exhausting day and you are facing an uphill challenge," says Natasha Daniels of AnxiousToddlers.com:

The most effective way to make bedtime less painful—because it will probably be painful on some level—is to try and make the room as similar as possible to your children's bedrooms at home. At the peak of one of my child's anxieties, I would pack her lamp with us! Luckily, it was one of those fiber-optic lights that expanded but folded up nice and neatly. . . . I would recommend packing their nightlight and bringing whatever music they listen to at home. You want the room to look as similar to their room in the dark; that way if they wake up in the middle of the night, the room will look somewhat familiar.[5]

"We bring favorite blankets or pillows, sometimes even twinkle lights," says Dana Marciniak, a parent from Buffalo, New York. Meanwhile, Brooklyn parent Delight Dolcine says she brings a mini vacuum when staying in vacation rentals, "because [my daughter] leaves dried up Play-Doh all over. I find the local grocery [store and] stock up on vinegar which gets out the slime stains."

Finally, whereas Sarah Salter recommends bringing child-friendly items like pump foam soap for the bathrooms, Kelli Engstrom, a parent from Kent, Washington, says that when her oldest was smaller, "we would put the ironing board in front of the door to help with elopement. Now we bring a Roku, so we have their calm-down television shows."

What's clear from these tips is that with a little forethought, research, and ingenuity, parents can turn the dream of a stay away from home into a reality. Next, I'll provide some hotels that have done some of that advance work for you, marketing themselves as autism friendly or autism certified. I'll also explain what to look for in a vacation rental, and then discuss how a resort like Beaches or a dude ranch accommodation might fill the bill. And in the next chapter, I'll cover another form of accommodations that's a travel subject in itself: camping.

AUTISM-FRIENDLY RESORTS

At the time of writing, the following had either been designated by IBC-CES as Certified or Advanced Certified Autism Centers (indicated by, respectively, CAC or ACAC), had been certified by the Center for Autism and Related Disorders (indicated by CARD), or had expressed their autism friendliness without specific designation (indicated by AF). Those properties that at press time were no longer included in the 2022 *Autism Adventure Guide* (https://autismtravel.com/autismadventureguide-signup-2022/) or the *Autism Travel Directory* (https://autismtravel.com/travel-directory/) are labeled as "2021 CAC."

Please research accommodations ahead of time to ensure they will work for your family, especially those marked AF because modifications can vary tremendously. Please also note that this is not an all-inclusive list, and as time goes on, some properties may become autism certified or autism friendly while others may drop off the list entirely. For the most current information, including additions to the list, be sure to couple the use of this book with visits to my blog, TravelingDifferent.com, and my Twitter feed at http://www.twitter.com/@Travelingdiff.

- Arizona:
  - **Saguaro Lake Guest Ranch** (2021 CAC)
    13020 N. Bush Hwy, Mesa, AZ 85215
    480-984-2194
    http://www.saguarolakeranch.com

  - **Courtyard Phoenix Mesa** (CAC)
    1221 South Westwood, Mesa, AZ 85210
    480-461-3000
    https://www.marriott.com/hotels/travel/phxme-courtyard
    -phoenix-mesa

  - **Delta Hotel Phoenix Mesa** (CAC)
    200 North Centennial Way, Mesa, AZ 85201
    480-898-8300
    https://www.marriott.com/hotels/travel/phxde-delta-hotels
    -phoenix-mesa

  - **Four Points by Sheraton at Phoenix Mesa Gateway Airport** (2021 CAC)
    6850 E. Williams Field Road, Mesa, AZ 85212
    480-579-2100
    https://www.marriott.com/hotels/travel/phxma-four-points-at
    -phoenix-mesa-gateway-airport

  - **Sheraton Mesa Hotel at Wrigleyville West** (2021 CAC)
    860 N. Riverview, Mesa, AZ 85201
    480-664-1221
    https://www.marriott.com/hotels/travel/phxww-sheraton
    -mesa-hotel-at-wrigleyville-west

- Florida
  - **Doubletree by Hilton Hotel Orlando at SeaWorld** (2021 CAC)
    10100 International Dr, Orlando, FL 32821
    407-352-1100
    https://www.hilton.com/en/hotels/mcosrdt-doubletree
    -orlando-at-seaworld

- Springhill Suites by Marriott Orlando at
  SeaWorld (2021 CAC)
  10801 International Dr, Orlando, FL 32821
  407-354-1176
  https://www.marriott.com/hotels/travel/mcoss-springhill
  -suites-orlando-at-seaworld

- Fairfield Inn and Suites by Marriott Orlando at
  SeaWorld (2021 CAC)
  10815 International Dr, Orlando, FL 32821
  407-354-1139
  https://www.marriott.com/hotels/travel/mcofw-fairfield-inn
  -and-suites-orlando-at-seaworld

- Tradewinds Island Resorts (CARD)
  5500 Gulf Blvd, St Petersburg Beach, FL 33706
  800-249-1667
  https://www.tradewindsresort.com

- Sawgrass Marriott Golf Resort and Spa (AF)
  1000 Tournament Players Club Blvd, Ponte Vedra Beach, FL
  32082
  904-285-7777
  https://www.marriott.com/hotels/travel/jaxsw-sawgrass
  -marriott-golf-resort-and-spa

- VillaKey Villa and Condo Rentals (AF)
  Main office: 1420 Celebration Blvd Suite 200, Celebration, FL
  34747
  407-569-2280
  https://myvillakey.com

- Texas
  - Wyndham Garden Hotel (AF—closed due to renovation
    until 9/29/22)
    3401 S IH 35, 1600 Woodward St, Austin, TX 78741
    512-448-2444
    https://www.wyndhamhotels.com/wyndham-garden/austin
    -texas/wyndham-garden-austin/overview

- **Omni Houston** (AF)
  4 Riverway, Houston, TX 77056
  713-871-8181
  https://www.omnihotels.com/hotels/houston

- Vermont
  - **Smugglers' Notch** (AF)
    4323 VT-108 South, Jeffersonville, VT 05464
    802-332-6854
    https://www.smuggs.com

- Wisconsin
  - **Don's Bay Getaway Airbnb** (AF)
    2 S. Walworth Avenue, Williams Bay, WI 53191
    https://www.airbnb.com/rooms/44195690?source_impression
    _id=p3_1595499742_7uOZarNATGCX5FtR
    Note: Partnering with the non-profit Sensory City organiza-
    tion, this is the first "Sensory Friendly Certified" accommo-
    dation in the Lake Geneva area, offering such considerations
    as weighted blankets, sensory kits containing Bose noise-
    canceling headphones and sensory toys, a sensory swing in the
    kids' room, bean bag chairs, essential oil diffusers, door alarms,
    etc. They use non-toxic/plant-based cleaning solutions and
    fragrance-free detergents for their linens.[6]

- Outside the United States
  - Belize:
    - **Ramon's Village Resort** (CAC)
      Coconut Drive, San Pedro, Ambergris Caye, Brazil
      U.S. Office: 800-624-4215
      https://ramons.com

  - Canada:
    - **The Chelsea Hotel, Toronto** (AF)
      33 Gerrard St W, Toronto, ON M5G 1Z4, Canada
      416-595-1975
      http://www.chelseatoronto.com

- **Hotel Port Aux Basques** (AF)
  2 Grand Bay Rd., Channel-Port aux Basques, Newfoundland A0M 1C0, Canada
  877-695-2171
  http://hotel-port-aux-basques.com

- ○ Dominican Republic
  - **Grand Palladium Bavaro Suites Resort & Spa** (AF)
    Avenida Francia s/n, Playas de Bavaro, Higüey 23000, Dominican Republic
    52-984-873-4825
    https://www.palladiumhotelgroup.com/en/hotels/republica dominicana/puntacana/grand-palladium-bavaro-resort-spa

  - **Nickelodeon Hotels & Resorts, Punta Cana** (AF)
    Carretera Uvero Alto Parcela, Punta Cana 23002, Dominican Republic
    809-468-0505
    https://www.karismahotels.com/nickelodeon-hotels-resorts /punta-cana

- ○ Jamaica
  - **Beaches Resorts, Negril** (ACAC)
    Long Bay, Negril, Jamaica, Norman Manley Blvd, Jamaica
    888-232-2437
    https://www.beaches.com/resorts/negril

  - **Beaches Resorts, Ocho Rios** (ACAC)
    Old Main Road, Ocho Rios, Jamaica
    888-232-2437
    https://www.beaches.com/resorts/ocho-rios

  - **Franklyn D. Resort & Spa** (AF)
    Main Street, Runaway Bay, St. Ann, Jamaica
    876-973-4124
    https://fdrholidays.com

- • Turks & Caicos
  - **Beaches Resorts** (ACAC)
    Lower Bight Rd, The Bight Settlement TKCA 1ZZ, Turks & Caicos Islands
    888-232-2437
    https://www.beaches.com/resorts/turks-caico

BEACHES RESORTS

Beaches Resorts has long been a leader in providing autism-friendly accommodations. In 2017, with three resorts—two in Jamaica and one in Turks & Caicos—it became the first resort company in the world to receive the CAC designation from the IBCCES and, in 2019, the first to elevate that designation to ACAC, which it has extended through 2023.

What makes a CAC or ACAC resort different than a typical property? For Beaches Resorts, it means advanced autism training for its staff to increase sensitivity—offered virtually during the recent pandemic—with an emphasis on key touch point areas, including kids' camps, entertainment, front desk/reception, food and beverage, and watersports operations. As part of its recertification, Beaches will also expand its training of team members to the airport arrival lounge and reception area in Montego Bay, Jamaica, and Providenciales, Turks & Caicos, which offer the first point of contact for Beaches' arriving guests.[7]

Janeen Herskovitz, a licensed mental health counselor who specializes in helping autism families, says she thinks it's great that more resorts are adopting an autism-friendly attitude:

> Families often feel very isolated, even when out in public. The possibility of a meltdown or unexpected behavior is always present. The fear of judgment from others is difficult to get past. Truly allowing our children to be themselves can only happen in an atmosphere of acceptance and inclusion. And this atmosphere can only be achieved when everyone involved is educated and willing to learn. Awareness doesn't just mean others know autism exists, it means they know what to do to help your child and family feel accepted, even when things don't go as planned.

At Beaches, families on the spectrum will benefit from:

1. In-depth education for Beaches Resorts staff, with forty credit hours on autism sensitivity and awareness, centered on communication, motor skills, social skills, environment awareness, emotional awareness, bullying, early childhood identification, transition to adulthood, and more.
2. Changes to protocols and physical space to accommodate various needs.

3. An optional service of "One-on-One Beaches Buddy" providing personalized, private childcare with a buddy who is autism-certified and can be prebooked for a nominal fee.

4. A dedicated toll-free number to Beaches' Special Services Team, certified by IBCCES, where Beaches staffers can assist both consumers and travel professionals who wish to learn more about a Beaches vacation and on-resort autism programming.

5. A Culinary Concierge program to support specific dietary restrictions and special requests.

6. Modified check-in options for private, in-room check-in.

7. Availability of sensory toys, as well as DreamPad pillows for children's use.

8. Modified design and decoration in Kids Camps and key entertainment areas to create a more sensory friendly environment.

Beaches also will launch a Sensory Stimulation Guide for guests with sensory needs, providing comfort and an understanding of what to expect in each designated area of the resort. The guide, which outlines the degree of sensory stimulation in a specific area or event, allows families to easily plan and navigate their visit based on their individual needs.

Designated Low Sensory Areas also will be identified at all resorts, allowing guests to find comfort in designated spaces should they need a break from sensory stimulation. These locations will be easily identifiable through resort maps, onsite signage, and in pre-travel planning materials.

"Like the rest of the world, we know families and autistic individuals are looking forward to traveling, visiting new places, and making new memories as soon as they are able to," says Myron Pincomb, IBCCES board chairman and chief executive officer. "They are also looking for organizations that are trained and certified in autism, particularly leaders like Beaches Resorts, who go above and beyond. The professionalism, dedication, and enthusiasm of the team at Beaches Resorts is second to none, and we are thrilled to continue our partnership to provide long-lasting support and impact."

In addition, Beaches Resorts' industry-leading Platinum Protocols of Cleanliness were designed with guests with sensory disorders in mind. Such protocols include the use of low-fragrance cleansers and fragrance-free hand sanitizers.[8]

For more information on Beaches' long-standing commitment to creating an autism-friendly environment, visit their website at Beaches.com /all-inclusive/autism-friendly.

VACATION HOUSE/APARTMENT/ CONDO RENTAL TIPS

Margalit Francus, who runs the website AutisticGlobetrotting (autistic globetrotting.com or https://www.facebook.com/autisticglobetrotting), offers the following tips for parents of children on the spectrum when evaluating possible villa and condo rentals:

- Things to research ahead of time:
    - Nearby grocery stores and drug stores, especially those that are open twenty-four/seven. The list should include driving distance from the vacation home as well as the stores' phone numbers.
    - Play areas and playgrounds with swings which are near the house.
    - Five sensory activities that are nearby to entertain the kids.
    - The name of a vet for those traveling with a service dog.
    - Nearby stores that offer gluten-free foods along with their phone numbers.
    - The distance of the house from the major theme parks with available transportation.

- It is best if there is more than one lock on the front door, preferably something which is harder for the child to open than just a regular deadbolt. If that's not possible, it's best to find a property that has door sensors with chimes on all the doors and windows.
- Pools should be enclosed by a [fence and accessed by a] gate with an alarm.
- Have chemical-free sheets, towels, and pillows on hand.
- Standing floor lamps are always a concern as well as full-size mirrors. Most autistic children are hyper and anything with glass is a major concern.
- The more familiar the child is with where they are going to be staying, the better. Ask the villa company or homeowner for at least

ten to fifteen pictures of the house, including the bedrooms, the kitchen, the pool area, and the living room.

- When choosing a condo or rental home, it's always good to choose one with a handheld shower since most parents must help their children with their baths. Also look for grab bars and non-slip bathmats.
- Creating a sensory area is a good idea; [perhaps] include a yoga ball, bubble wrap boards, and synthetic pieces of grass. Having a game play system on hand is advised, and at least one room in the house should have lighting that can be dimmed.[9]

DUDE RANCH ACCOMMODATIONS

For children who receive equine therapy at home, a dude ranch stay might be a great segue into a family vacation. Rupert Isaacson has written two books, *The Horse Boy* and *The Long Ride Home*, where he describes the therapeutic effect of his visits to Mongolia, Namibia, and Navajo reservations with his teenage son. The curative properties of equine interaction might also aid your family closer to home. While dude ranches themselves might not have autism-friendly modifications in place, the opportunity to commune with the horses could be healing in itself, which might make them the perfect accommodations choice for a horse-loving neurodiverse traveler.

Gene Kilgore of Top50Ranches.com, in his 2015 article, "Horseback Riding for Kids with Autism," writes:

> One common character trait of children with autism is that they often experience problems in bonding with other people because of their difficulty with traditional methods of communication. Many autistic children struggle with verbal development and eye contact and other ways that people use to bond with each other. With equine therapy, since the children are working with horses who interact in a purely physical manner, they can learn other ways to connect, create, and nurture bonds. . . . [The therapy] can be very soothing for autistic children who may experience high anxiety or frequent tantrums.[10]

A few dude ranches that either advertise themselves or declare themselves as autism friendly include:

- Arizona
  - **White Stallion Ranch**
    9251 W Twin Peaks Rd, Tucson, AZ 85743
    520-297-0252
    https://www.whitestallion.com

  - **Tanque Verde Ranch**
    14301 E Speedway Blvd, Tucson, AZ 85748
    520-296-6275
    https://www.tanqueverderanch.com

- California
  - **Coffee Creek Ranch**
    4310 Coffee Creek Rd, Trinity Center, CA 96091
    530-266-3343
    https://www.coffeecreekranch.com

- Colorado
  - **YMCA of the Rockies Snow Mountain Ranch**
    1101 Co Rd 53, Granby, CO 80446
    970-887-2152
    https://ymcarockies.org/Locations/Snow-Mountain-Ranch

- Montana
  - **The Ranch at Rock Creek**
    79 Carriage House Ln, Philipsburg, MT 59858
    877-786-1545
    https://www.theranchatrockcreek.com

  - **Erik's Ranch**
    273 Trail Creek Rd, Livingston, MT 59047
    (612) 401-3080
    https://eriks-ranch-montana.business.site

- New Mexico
  - **Geronimo Trail Guest Ranch**
    1 Wall Lake Road, Winston, NM 87943
    575-772-5157
    https://geronimoranch.com

- Texas
  - **Dixie Dude Ranch**
    833 Dixie Dude Ranch Rd, Bandera, TX 78003
    830-796-7771
    http://www.dixieduderanch.com
- Wyoming
  - **Medicine Bow Lodge**
    5556 Hwy 130, Saratoga, WY 82331
    307-326-5439
    https://medicinebowlodge.net
- Outside the United States
  - Mexico
    - **Rancho Las Cascadas**
      San Agustin Buenavista, San Francisco Soyaniquilpan de Juarez, C.P. 54280, Estado de Mexico / Mexico
      52-1-55-10-70-20-80
      https://www.rancholascascadas.com

For a comprehensive listing of all dude ranches, autism friendly or not, contact ranching expert Gene Kilgore at Top50Ranches.com or the Dude Ranchers' Association at https://duderanch.org/find-a-ranch.

PARENT SPOTLIGHT—
KRISTEN CHAMBLISS, LEAGUE CITY, TEXAS

Kristen Chambliss is a parent with two children, one neurotypical and the other neurodiverse with ASD and auditory processing disorder. She's traveled with her daughter with ASD since the child was four months old "with no apprehensions" since she wasn't diagnosed until age four. "My daughter's an easy traveler. Because she's used to it, she usually does great," she says.

"Usually" is the operative word here. When things did not go as planned on a trip to the Caribbean, the resulting situation required some quick thinking and adaptation on Chambliss's part:

> We went to Beaches in Jamaica, which was probably one of our best trips ever. However, when the [life-sized Sesame Street] characters

popped up unexpectedly, it totally freaked out our ASD kiddo, and when the younger one saw her freak out, she freaked out as well. We decided it was their vacation too . . . we figured out where the characters went and when, and then avoided them.

Another lesson learned by Chambliss is that for children with ASD, previewing trip endings is as important as trip beginnings. She recounts:

On that same trip, I forgot to tell my ASD kiddo that we were going home after Jamaica, so when we arrived home at midnight (the plane was majorly delayed), she was upset and insisted she did not want to be home and wanted to go back to the hotel in Jamaica. We did end up getting her to sleep, but only after about an hour of crying. It was pretty exhausting, but she was five, and was fine the next day.

Chambliss says she usually plans travel on her own, though she used a travel agent for Disney World at a time when she had no special needs requirements. "We travel because we want to expose our children to the world. We've gone to other states and countries, sometimes for work, sometimes relaxation. We always try to find something fun for them to do."

CHAPTER 12 ACTION STEPS

1. To select accommodations, rely on the personal experience of fellow parents with children on the spectrum, or a specialized travel professional rather than internet descriptions whenever possible.
2. Select an autism-friendly or autism-certified accommodation to ensure your special needs will be top of mind with the property. Many such hotels and resorts are listed in this chapter.
3. Prepare ahead with Social Stories and videos.
4. Decide if a hotel or a vacation rental might be a better choice for your family. Each has its pros and cons. Study the included checklist of what to look for or request when renting a vacation apartment, condo, or house.
5. Research the area surrounding the property to ensure it has the amenities you'll need. Decide in advance what amenities are musts, such as a pool, an in-room kitchen or kitchenette, no balcony, etc.

6. Choose your room's location and size based on your child. Do you need the space two adjoining rooms would provide? Do you need an upper floor or would a lower one be more convenient?

7. If the room doesn't have a bathtub and that's what you need, bring along a small, blowup pool and place it in the bottom of the shower.

8. Bring fans if you need to drown out noise from hotel hallways.

9. Make sure the room has Wi-Fi if that's what you rely on to keep your child distracted.

10. Check in without your child or choose remote check-in if lobby noise and crowds will present a problem.

11. Make sure children have a "retreat"; perhaps a hotel closet will do. Structured activities are also a must for the child who has difficulty making friends.

12. Be flexible with sleeping times (extra naps, earlier bedtimes) once you arrive.

13. Request special cleaning if your child has allergies.

14. Bring along from home any items that will make the room familiar and safe.

15. Consider a dude ranch as an accommodation if your child feels more comfortable relating to horses than people.

13

THE GREAT OUTDOORS
Camping, RVs, and More

"We had a hard time with our older, neurotypical son on our camping trips—he wanted to spend the whole first day in the tent, reading, but we felt like we were wasting precious time. We decided from then on, we'd just bump out our trips one extra day so he could have a chance to adjust, and we didn't feel the need to rush him into something he didn't want to do.

For our younger son with autism, we keep physical activity light on the first day. He has low muscle tone and tires easily, especially after a long day of travel. We call that first day our 'entry day,' like we've just landed a spacecraft and need time to decompress." – Adrienne Query-Fiss, parent, Seattle, Washington

For those families on the spectrum who are still wary of air travel or cruising, camping may be a great, affordable vacation. You can experiment with being away from home without venturing too far afield and while limiting encounters with strangers. You can also stick somewhat to your basic, at-home routine in terms of menu and activities.

Your biggest camping decision might be whether to sleep in a cabin, a tent, a pop-up camper, or a recreational vehicle (RV). Your child's particular spatial needs as well as where you fall on the roughing-it-to-luxury-resort continuum may be a determining factor. Luckily, many campgrounds have websites or videos to allow you to preview what's to come, and glamping or glamorous camping (Glamping.com)—which includes resort-style services not usually associated with "traditional" camping—has come into vogue.

Tent camping is a good way to get outdoors and enjoy fresh air and a new destination in a controlled environment, says Jonathan Haraty of Adventure Luxury Travel in Whitman, Massachusetts, but you should look for campgrounds where the individual sites are spread out so that you do not have partying adults five feet away from you. And look for any other triggers that might set off your son or daughter. Think about how they [might] react to animals wandering around, using communal bathrooms, and in general, just being away from their own bed for multiple nights.

"On the other hand, an RV gives you an even more controlled environment and works great with young adults on the spectrum who have difficulty dealing with crowds," he says. "There is no need to set up a tent or pop-up camper. You have your own bathroom, and you are pretty much self-contained; you can control your surroundings more than by tenting, but it will be a more expensive vacation."

"RV vacations give you a lot more control over sensory stimuli and your schedule than a hotel stay might," says Jennifer Hardy of Cruise Planners in Kent, Washington:

> You can also visit areas that are a little more remote, giving your child the freedom to move and explore more comfortably. Many people do not realize that travel advisors can help you with smaller vacations like camping, road trips, and RV rentals! Using a travel advisor for these excursions gives you an expert in your corner and a resource to help you plan an itinerary that will match your family's interests and needs.

With RV travel, "make sure there are plenty of items to keep kids occupied while mom and dad are driving," suggests Sarah Salter of Travelmation dba Sarah Travels Happy in Fayetteville, Georgia.

Just because you start in an RV doesn't mean you have to end in one, assures Carl W. Johnson of Adventures in Golf (incorporated as Hamill Travel Services, Inc.) in Amherst, New Hampshire. As a family with ultrasensitive members who might have issues with tight confines, "you are entirely in control—you can stop when you want, you can even rent an Airbnb or hotel room along the way if lack of space becomes an issue. You truly are operating your own holiday and can likely sidestep any possible issues you might encounter. Of course, you are limited as to where you can travel, but you ultimately have full control over the entire trip."

Claustrophobia aside, there are other issues to be contemplated before renting an RV or setting up a tent.

"Is the child familiar with outdoor smells, terrain, sounds, sights?" asks Betty J. Burger, recently retired from BGB Travel and Leisure, LLC in St. Marys, Georgia. These are not inconsequential questions for young campers on the spectrum. In YourAutismGuide.com's *2019 Guide to Camping*, Angela Zizak describes the potential sensory assault of campgrounds, such as

> the smell of the campfire and fuels from cars driving by, and the pungent smell of the closest outhouse. The sight of the fire lighting up the night, or games and other social activity between people at different campsites. The sounds of nearby site conversations, kids yelling to each other on their bikes; loud diesel trucks going by; birds and other animals; dogs barking, etc. New tactile sensations like camping pads made up of gravel rocks with some dirt or grass, or others made of concrete. Handling wood or sitting on a rough picnic table though, if your spectrum child likes deep pressure, swing hammocks are a perfect way to relax. And the taste of different waters unless you bring your own.[1]

To introduce some of these unique sensory experiences to the child, it might be advisable to practice an outdoor overnight trip before embarking on a longer camping vacation. You can look for a local campground and try out the experience for just one night to see how it goes. But, adds Haraty, "[like] any other experience for your spectrum child, prepare them for it first by going over what they are going to see, hear, and feel on this outdoor adventure."

If renting an RV is out of your budget, and you want more protection for a "wanderer" than the zipper on a tent can provide, a camping cabin might be a consideration. They offer more of the comforts your child expects at home, including a proper bed and even a private shower and toilet, and being well-insulated, they're better than tents at keeping sounds and the weather outside. Wigwam® Holidays in the United Kingdom assures travelers that their cabin doors are also lockable for greater peace of mind. With more than eighty sites available nationwide across farms, coastlines, and in forests across England, Scotland, Wales, and Northern Ireland, they say their aim is to help travelers find a campsite suitable to their family's unique needs. Many of their sites are dog friendly and some offer wheelchair-accessible cabins specially designed for those with additional mobility needs.[2]

Also in the United Kingdom, Leafy Fields has earned an "autism-friendly" award from the National Autistic Society, among other honors.

Located in rural Devon, owners Andrew and Dannie Sheard describe their venue as a family-friendly glamping site with

> a shepherd hut, two safari tents and a bell tent. We have a variety of animals including mini horses, chickens, guinea pigs and mini sheep, as well as a willow garden designed for us by Alan Gardner, aka "the Autistic Gardener." We offer a Sunflower Pack for those with hidden disabilities including autism. It contains social stories, [suggestions of] places to visit during a stay, a sunflower lanyard, and animal communication cards. We also have a sunflower sensory playroom available for guests.

Some campgrounds closer to home are geared specifically to neurodiverse travelers. One such campground is Camp Yofi, which describes itself as "a one-of-a-kind program for families with children with autism spectrum disorder between the ages of six and thirteen."[3] Run by the Ramah Darom organization (though open to all, regardless of denomination) and located in rural North Georgia, Camp Yofi invites a maximum of twenty-five families, which can include parents, grandparents, and siblings accompanying the children on the spectrum. Each population gets a separate activity track during the day and then everyone comes together in the evening for campfires, guitar sing-alongs, s'mores, and other family programs. Each Camp Yofi family receives a *chaver* (a special friend) who spends mornings and evenings with the same family throughout the week, providing a helpful, supportive, and consistent presence.[4]

If you're not traveling to a campground specifically designed for families on the spectrum, here are a few other things to be aware of as you tackle the great outdoors.

"It's important to plan out your route carefully before departure," says Jennifer Hardy of Cruise Planners in Kent, Washington. "You will need to know in advance which stores on your route carry your child's favorite foods or daily living supplies so that you never run out. You will also want to make reservations for your overnight parking spots in advance. Whether you want to be close to the restrooms or in a more remote location on the property away from noises and lights, where you park your RV overnight can make or break your experience."

Angela Romans of Rejuvia Travel in South Bend, Indiana, agrees that the less you leave to chance, the better. Her list of questions includes: "What features/amenities will be available? Will you have direct access to water/electricity? Will you be camped in the sunlight or the shade, because this can make a difference to someone with sensitivities to heat or sunlight. How much privacy will you have?"

Lastly, she cautions neurodiverse families to make absolutely sure to stay far from the water if their child can't swim, because drowning is the leading cause of death of children with autism. They are 160 times more likely to die from drowning than the general population of children.[5]

Romans explains: "Individuals on the spectrum can wander or attempt to 'escape' situations unpredictably. Many believe the soothing qualities of water, along with its tendency to reflect, shimmer, and produce patterns as it moves, is what draws people to it. Know in advance where you will go for assistance if your child wanders off," she says.

"Many areas of worry come along with tent camping no matter what your child's diagnosis or lack of same," warns Lisa Bertuccio of Vacation Your World, Otisville, New York. "The lack of bathroom facilities, open fire, an unsecure exit/entrance, to name a few. Be sure to go over safety rules many times prior to the trip. If you go tent camping, choose a campsite that provides bathroom facilities. And make sure your child has an identity bracelet in case he or she wanders off." Another safety measure, offered by Brenda Revere of C2C Escapes Travel in Bedford, Indiana, is to sew a GPS unit into children's clothing in case they unexpectedly stray.

To make the vacation a memorable one, Nicole Graham Wallace of Trendy Travels by Nicole—Dream Vacations in McDonough, Georgia, advises camping clients to choose a location that provides kid-friendly activities. "Make sure to have all their favorite toys, blanket, and entertainment as this will be key to having a pleasant experience," she says.

You can also enhance the adventure by promoting the novelty of this mode of travel. Michelle Zeihr, a parent from Barrie, Ontario, Canada, suggests bringing along a cool sleeping bag and letting kids cook hot dogs on a stick. Meanwhile, Kristen Chambliss, a parent from League City, Texas, takes advantage of emotional connections with fictional characters her child knows and trusts to prepare her for the vacation. "She loves Peppa Pig, so we told her we were getting a camper van like Peppa Pig's, and she was ecstatic for days."

At the campsite, have a set of camping rules; kids do better when they have structure, says blogger Tabitha Philen in her article, "Camping with Special Needs Kids." "If these rules are written out, kids will be able to refer to them (if they can read). You can have a family meeting to discuss them. This will help the kids to know what they can and cannot do. And make safety a part of the fun and get everyone involved. Give your kids different scenarios and ask them what they should do if something happened."

"Assign different jobs to everyone," she continues:

> This will help you divide the work and responsibilities. Be sure each person knows what they are in charge of. Kids might grumble at the mention of chores but it you emphasize teamwork they are more willing to assume responsibility. Sometimes extra adult supervision is required to keep everyone safe. If needed, recruit other family members or family friends to come along for additional support. And place everyone on the buddy system so no one gets lost but rotate [buddies] to keep everyone happy and balanced.

Philen also advises parents to consider purchasing items that can make their camping trip easier, including locks, headlamps, activity books, telescoping roasting sticks for safer s'mores making, glowing light sticks, battery-powered LED light bulbs, inflatable air mattresses, weighted blankets, and more.[6]

By following these strategies and precautions, camping can be a safe and memorable vacation for everyone and the first step to longer and more varied holidays. What follows is a helpful questionnaire to help evaluate campsites, some information on free park passes for families on the spectrum, and details about an exciting activity, geocaching, that you can sample while away.

CAMPGROUND QUESTIONNAIRE

Here is a list of questions to consider when booking a campground which first appeared in YourAutismGuide.com's *2019 Guide to Camping*, reprinted with permission from author Angela Zizak:

- Is the campground and/or some sites completely in a wooded area or in full sun? This can be a critical element if someone in your family is sensitive to sun and heat.
- Is there natural privacy and plenty of space between sites, or is there very little privacy?
- What is the terrain like? Rough, steep, or flat?

- Does it have access to water and electricity, or will you be totally roughing it?
- Does the campground and site fully accommodate someone who needs a wheelchair or other medical equipment?
- Is there Wi-Fi available in the case of emergency (or to keep your little ones amused)? Or can you easily reach someone at the campground office? Many state and national parks have limited or no internet access, while private campgrounds may have Wi-Fi available for guests.
- Is your site accessible to other activities and restaurants, or is it remote?
- Are pets allowed? What rules are in place if dogs are allowed?
- If your child is a runner, will you be able to get help right away?
- Is the location near water? This is especially important to consider if your child is not yet a swimmer.
- Would you feel comfortable giving your child some room to explore the campground on his or her own?
- What amenities will be available?
- Does the cost of the campsite and rentals fall within your budget?
- How new are you to camping? If you have never been inside a tent or camper, rather than buying one, consider renting a camper from a company like Cruise America. Some campgrounds will even have their own campers available for you to rent. Others, like Fort Wilderness at Walt Disney World, will allow you to rent a camper that is brought in and set up by a local company. You get the benefits of camping without all the work.[7]

AMERICA THE BEAUTIFUL ACCESS PASS

The America the Beautiful National Parks and Federal Recreation Lands Access Pass is a national park pass for those with autism. It gives you access to locations like the Grand Canyon, Yellowstone Park, and the Smoky Mountains free of charge, as well as deep discounts on camping and other recreational activities.

According to the application, the Access Pass is a free, lifetime pass available to U.S. citizens or permanent residents, regardless of age, who have a permanent disability. Applicants will be required to provide documentation of permanent disability and residency or citizenship. The pass can be used at over two thousand federal recreation sites across the nation, including national parks, national wildlife refuges, and many national forests and other federal recreation lands. It admits the pass owner and any passengers traveling with him/her in a single non-commercial vehicle at per-vehicle fee areas, or the pass owner and three additional adults where per-person fees are charged. The Access Pass may also offer a discount on some expanded amenity fees, such as camping. Discounts offered by the pass vary widely across the many different types of recreation sites. Before ordering one, you should check with the recreation sites you plan to visit to confirm that the pass will be accepted. Anytime a pass is used, photo identification will be requested to verify ownership.[8]

To learn more and apply, go to www.nps.gov/planyourvisit/passes .htm. You can learn more about camping destinations at www.recreation .gov as well as at the National Park Service site at https://www.nps.gov/ index.htm.

GEOCACHING WHILE CAMPING

Geocaching is not only an activity that's easy to take on the road, it could become the highlight of your camping trip as well as future excursions. According to a 2019 article on The Good Men Project by J. R. Reed, an author, adult autism advocate, and self-proclaimed "proud Aspie" (a person with Asperger's syndrome), this real-life scavenger hunt has attracted fans worldwide. Simply put, people hide things, called caches, and upload information and coordinates for the cache. Then, using a handheld GPS device or simply a smartphone with the geocaching app, other people navigate their way to the cache of interest.

Inside a typical cache will be a log you'll sign and date, proving that you found it. Many have little trinkets inside that you can trade, which is great for collectors. Most geocachers take what's inside and leave something behind of similar value.[9]

So why does Reed say that this has particular appeal for those with autism spectrum disorder and social anxiety disorder?

If you're on the spectrum and need to get out of the house and do something inexpensive, I wholeheartedly recommend geocaching. [While] I don't do a lot of things with other people, this gets me . . . driving around the Ozarks with my service dog, Tye. Sometimes I forget how good it feels just to get out, no matter what the weather. It's also a fun family adventure, especially if you're taking a road trip and you have a little time to kill along the way.

Erin Clemens, who also self-identifies as an adult on the spectrum, adds:

When I was a child, I felt like I didn't belong anywhere. Through activities such as geocaching, I finally feel a connection to the community. And while I may not be traveling the world, I'm driving further than I thought I could because of the containers waiting for me just outside of my comfort zone. With geocaching, I'm not just playing around. I'm finding ways to grow. And that's the best hidden treasure I could ask for![10]

And in a 2010 story in the *Oregonian*, Matt Buxton described how geocaching provided a way for "painfully shy" then nine-year-old Ryan Hurley, whose autism severely affected his speech, to interact and communicate with his family. Says his mother, Beth: "It's like his interest in finding treasures overrides his need for routine."

"Ryan's enthusiasm isn't surprising," says Dr. Darryn Sikora, a clinical psychologist in Oregon, who explains that many people who have autism often focus their interest on one or two specific activities and thus, are able to communicate much better. "It's fairly common for individuals with autism to have better social communication around their intense areas of interest," she says. "Unfortunately, in the twenty-first century, many of the individuals have latched on to video games, and it's difficult to have social interaction and communication while playing a video game."[11]

While there are many places to learn about geocaching, Reed recommends the site Geocaching.com to read more and download a free app for your phone with a built-in compass and GPS. He adds, "Unless you don't find a cache that you searched for, which hasn't happened to me yet, you know there will be a victory at the end. And a victory is definitely worth celebrating."

PARENT SPOTLIGHT—
ADRIENNE QUERY-FISS, SEATTLE, WASHINGTON

Adrienne Query-Fiss is the mother of two, one on the spectrum with autism spectrum disorder and attention-deficit/hyperactivity disorder. She's been traveling with them since the youngest was five weeks old, but she says it hasn't always been easy: "My autistic six-year-old can be disruptive when he's anxious or upset. He's a big boy for his age and I worry he will hurt himself or someone else."

"We generally like to do things we all enjoy, whether it's camping, hiking, or [visiting] a theme park. We all like an occasional change of scenery. Our main challenge is air travel—our autistic son gets claustrophobic on planes." However, she dismisses the misconception that travel with those on the spectrum is "impossible," especially because "shockingly, my autistic child has a much greater sense of adventure than my neurotypical child and is always up for a new experience. My NT kid is an introvert and a homebody and isn't always excited to take a trip."

Query-Fiss says she prepares both children the same way for a vacation, whether they're traveling domestic or abroad:

> There's lots of frontloading the experience (with visuals when possible), reading about where we're headed, making a plan for each day, and deciding together what we might eat when we're traveling. . . . Playing "What If We Went . . ." as a family activity has taken a lot of the anxiety out of travel for my kids. We also give our children some agency in what we decide to do, letting them choose one thing each to make the trip special—a souvenir, a special meal, or some one-on-one time with a parent.

CHAPTER 13 ACTION STEPS

1. Decide how you want to camp: tent, cabin, pop-up camper, or RV. Each has its advantages. Perhaps glamping (luxury camping) is more your style.
2. Preview your campground via website and video. Be aware of the possible sensory issues you may encounter.
3. Start small. You can even camp out in your backyard or a local campground for a night to start.

4. If you start in an RV, make sure there are a lot of distractions for your child. But if it becomes claustrophobic, you can break up the holiday and stop in a hotel or Airbnb. The beauty is that you're in charge.

5. Use the questionnaire in this chapter to know what questions to ask before booking a site.

6. With tent camping, choose a campground that provides bathroom facilities. And steer clear of sites near water if your child cannot swim.

7. Safety first: make sure your child wears an identity bracelet or sew a GPS locator into his or her clothes. Emphasize safety rules before departure and throughout the trip.

8. To keep kids excited, play on the novelty that comes with camping, from unusual sleeping bags to campfire cookouts.

9. Have a set of camping rules and chores for children and recruit additional family members when necessary for support.

10. Take advantage of freebees and deep discounts available to camping families with neurodiverse members.

11. Try geocaching to keep kids engaged and give them their own mission while away.

14

TO SEE OR NOT TO SEE

Tours and Venues

"Each family member should choose one activity they'd enjoy for the following day. Then, instead of parents saying, 'We're not going to do what you want until you stop whining,' recognize that they don't know how to soothe themselves and they are actually communicating with you, albeit in a mode you abhor. You can explain: 'It's okay for you to think this store is no fun. Later, we're going to see the movie you chose.'" – Dr. Ellen Littman, New York–based clinical psychologist

By air, car, bus, train, or ship, you've reached your destination. You've done your research and found the perfect, autism-friendly accommodations. You're set—except you can't spend all your time in the room, cabin, tent, or recreational vehicle. Part of the fun of travel is exploring each new destination. The challenge is figuring out how to do that without furthering the upset that a change in schedule may cause an anxious, inflexible, or neurodiverse young traveler.

"This is one part that I think many families find overwhelming," says Olivia McKerley, of Elite Events & Tickets in Grovetown, Georgia. "How will their child do with the crowds? How will they handle the change of routine? How often should they take breaks? Where are the places to take their child when they are overstimulated? These are all important considerations and can seem daunting. That's why working with a [Certified Autism Travel Professional] is especially important."

"We've found that places that offer low-sensory environments help," says Kristen Chambliss, a parent from League City, Texas:

For example, NASA is close to our home in Houston, and they offer low-sensory events with [limited attendance] on Saturday mornings for kids on the spectrum. It has lower lighting/natural light, fewer people, and is way less overwhelming in terms of noise. Sensory rooms like that settle kids and can help if they have meltdowns or struggle with anxiety. One of our daughters is neurotypical, the other has ASD, and we do a lot of things with both to calm them down since they have big emotions. Sometimes that means practicing kids yoga (using the GoNoodle app on our phones) and taking deep breaths for emotion regulation.

Talking to children before traveling, or rehearsing the trip ahead of time, can reveal wonders. For example, the beach, typically a tourist favorite, could be an issue for an autistic individual if he or she has hypersensitive sensory problems, writes Hend M. Hamed in "Tourism and Autism: An Initiative Study for How Travel Companies Can Plan Tourism Trips for Autistic People." He suggests that travelers with autism spectrum disorder (ASD) might have trouble tolerating changes in their environment, such as the crash of the waves, the sensation of sand particles, the novelty of the weather, and constant beating down of the sun on their bodies.[1] Perhaps experimenting with the feel of some sand bought at a local crafts store, and an hour or two spent at the shore—if you're local—could help parents avoid many vacation issues later on.

"Disney does not need to be a rite of passage for every family," says Starr Wlodarski of Cruise Planners—Unique Family Adventures in Perrysburg, Ohio. "And if the family members are sit-on-the-beach people, sending them to New York City to go to museums might not be the best idea." Wlodarski says this is where using a travel agent who knows you well helps, and also why getting your kids involved in the planning is important.

Though not right for every family, luckily many theme parks are autism-friendly, and their employees are trained to respond to children on the spectrum with compassion and skill. Many also have early entry and/ or special entrances.

"You or your travel advisor should vet venues prior to arrival," advises Sarah Salter of Travelmation dba Sarah Travels Happy in Fayetteville, Georgia. "Does the venue have special access lines, wheelchair rental, or diverse food options? Remind your children that they should look with their eyes, not their hands. If they need tactile experiences, prepare for that by bringing something along to keep their hands busy while waiting. Social Stories are huge here."

For success while touring, the onus is on parents to set a reasonable pace, as shown by the experience of Nyack, New York–based parent Sarah Ludwig:

> We have learned—and relearned again and again and again—that we can't force my ASD son to keep up the pace we expect. He has a lot of difficulty on travel days, so even if we are arriving at a destination early, we can't plan to do anything later. I am a planner and a list maker, and I like to see and do everything, so I very much need to slow down. Otherwise, it is not great for anyone. I need to remember to offer him lots of opportunities to rest and recharge. We went to Disney a few years ago, and he spent two or three hours every afternoon asleep in a carrier on my back (he was nine!) because he just couldn't handle the crowds and the heat and the pace. . . . I have many, many pictures of him sleeping through vacation because of the exhaustion of being outside his routine. So, the lesson is, manage expectations. Focus on one or two exhibits and then call it a day.

One way to decide on those exhibits is to virtually visit museums or amusement parks in advance, says Angela Romans, Rejuvia Travel, South Bend, Indiana. "Their websites can sometimes have social narratives with tips to prep parents for a visit, offer sensory friendly and interactive maps, and admission stickers/wristbands for reentry," she says.

Those interviewed offered several strategies for touring and conquering theme parks and other venues as follows:

GETTING HELP AND AVOIDING OVERSTIMULATION

Identify yourself as having someone in your party with special needs and determine what resources the destination may have to offer, advises Sandra Chu of The Enchanted Traveler in Rockford, Illinois. "For example, Disney World has a Disability Access Service (DAS) card which provides great flexibility for families with special needs. Provide the destination with the information they need to accommodate you best." (More on Disney later in the chapter.)

Because large crowds can be a source of anxiety for children with ASD, it is important to locate quieter places like nursing rooms or family rooms where the child can take a break from all the stimulation, advises Lisa Bertuccio of Vacation Your World in Otisville, New York. "For shows, be aware of where you can exit the theatre quickly to offer your child a sensory timeout if needed," she says. You can also seek out special days or times of day at theme parks that are either off-peak or specified as autism friendly, and therefore less crowded, adds Tara Woodbury of Escape into Travel in Leland, North Carolina. Jennifer Hardy of Cruise Planners in Kent, Washington, advises you to

> ask at Guest Services if any attractions or exhibits should be avoided with your child's sensory profile. Certified Autism Centers will have a detailed description of each venue and a recommendation for the type of individual who may be most impacted. Guest Services will also be able to set you up with any resources they may offer, like a sensory kit or an access pass.

Many locations now offer alternatives for children or adults who are unable to wait in traditional lines due to their disabilities, Hardy notes. "See if your destination will allow you to use your stroller as a wheelchair. This is a great option for younger children who are at risk of eloping. A wrist link or safety harness is another resource to keep your child at a more manageable distance," she says.

Gauge when your child has had enough and stop the activities for the day in favor of something calm and soothing, suggests Taylor Covington of *The Zebra*. "Even if this means you do not get to 'do it all,' you will still be creating positive experiences for your family, and your child will remain calmer as a result."[2] For many families, that means limiting "fun time" to a maximum of four hours for the child who needs "downtime," says Brenda Revere of C2C Escapes Travel, Bedford, Indiana.

For some travelers, Sarah Marshall of TravelAble in Midlothian, Virginia, recommends hiring an aide or a babysitter for these venues. "Especially at places like Disney, it can be very overwhelming and maybe you need to split up. There are many companies that offer help; they will accompany you into the park and look after your child if they need additional care," she says.

Those companies include Theme Park Nannies, which you can contact at Parknannies.com. Another is Extra Hands at Vacationsitterfl.com. Make sure to ask for a helper who has experience with children who are anxious, inflexible, or on the spectrum.

And keep safety first and foremost. Sarah Salter suggests writing your child's name, your name, and phone number with a Sharpie on his or her arm, or on a Magic Band (a product sold at Disney), or on clothing. "If you write on their arm, then cover what you've written with Liquid Bandage, so it won't come off in case of bathing, rain, or sweat," she says. Alternatively, Angela Romans recommends attaching a GPS tracker to a child who tends to run off.

ABOUT SIGHTSEEING TOURS

Keeping kids focused during a tour that's probably geared toward adults would be trying for any parent, let alone those with children on the spectrum. Part of the challenge is caused by the proximity of fellow tourgoers who may be crowding kids or talking behind them so they cannot hear the tour guide. One recommendation from Jonathan Haraty of Adventure Luxury Travel in Whitman, Massachusetts, is to try and find a tour that uses headsets to hear their narrative. You can also let the guide know your child is on the spectrum and what their triggers are. The guide may have some ideas on how to better help your child avoid an episode. Another idea, but of course more expensive, is to hire a personal tour guide or arrange for a private tour. "This works well in museums and some historic sites," he says.

That was Michael Sokol's experience. "A few times while traveling we hired a local guide to give us a private tour. Makes for a more relaxing adventure and when we were done, we were done," says the parent from Longmeadow, Massachusetts.

Another option is to book tours that offer priority access to reduce crowd size, says Suzanne Polak of AAA in Southington, Connecticut. Then, notify tour operators in advance about your needs.

Be sure to let the guide know if your child has specific areas of interest, as they may be able to add corresponding elements to your tour to make it a little more engaging for your son or daughter, adds Jennifer Hardy. "It is easier for a tour operator to identify which guests have mobility needs as opposed to sensory needs, so advance communication is important."

Group sightseeing didn't work for Michelle Zeihr, a parent from Barrie, Ontario, Canada:

Most of the time we had to ditch the tours and just go to places ourselves and check them out on our own. It's not always easy to adhere
to the schedule of tour operators. When you have a fussy baby, it can
be annoying, but people get it; often a fussy child is not looked at the
same way. We visited many ruins, museums, parks, etc., by renting a
car and going on our own. Just make sure you know [in advance] what
is around, such as parking, food, etc.

Delight Dolcine has a specific strategy to booking tours. Says the
Brooklyn-based parent:

> I always book from a third-party site like Viator or Groupon in case I
> need to file a complaint [later]. When we're in the Caribbean, I book
> air-conditioned coach charter buses. I order tickets in advance to avoid
> crowds and lines. I always go online and map out the bathrooms, res-
> taurants, and the Guest Services desk. [You can usually] go to Guest
> Services with the family member and get a band that lets you wait for
> less time or not at all. I don't go to places that don't make these reason-
> able accommodations.

For those traveling abroad, the National Autistic Society, based in
the United Kingdom, hosts a directory of companies in Europe that are
ready to tailor tours and accommodations to your spectrum needs, such as
HappyKidsHolidays.com for tours to France and Portugal4AllSenses.pt for
Portugal. See https://autism.org.uk/directory to access this directory.

MUSEUMS AND OTHER VENUES

Museums are popular tourist attractions and often take center stage on family travel itineraries. But Tara Woodbury warns that those on the spectrum
might find the crowds overwhelming. Another challenge might be movies
or other "interactive exhibits" that play music or sounds and have video
or flashing lights that come on automatically. "Your first order of business
should be to download a map of the museum to plan what areas might be
best avoided or to plan ahead with Social Stories," she says.

Luckily for neurodiverse families, hundreds of museums around the
country now cater to visitors with autism. The Smithsonian Institution in
Washington, DC, was one of the leaders in this movement. These offerings
vary from monthly get-togethers for families on the spectrum to explore

the museum during special hours to providing guests with noise-canceling headphones and weighted vests. Some have created sensory friendly rooms with warm lighting and interactive activities featuring tactilely inviting objects. Claire Madge is spearheading a movement to make more museums autism-friendly through her U.K.-based organization, Autism in Museums (AutisminMuseums.com), while Belinda S. Stull of Travel Leaders/ All About Travel, LLC, in Hagerstown, Maryland, points to the Georgia Aquarium as a great example of a museum that already offers a weekly autism-friendly program.

While there isn't a comprehensive list of ASD-friendly museums available, many are listed among the other types of venues later in this chapter. Be sure to do your own research when planning your trip and inquire if such programs are available. Something that might help with museums that lack a designation: former Reuters exec Topher Wurts has developed the "Autism Village" app, which allows people in the autism community to add, rate, and review different restaurants, museums, parks, playgrounds, and other locations based on "autism friendliness," the level of comfort or accommodation a venue is able to give a person with autism. Download it from http://about.autismvillage.com/app if you'd like to keep updated.

Also note that while not designated as autism friendly, many children's museums might be a wonderful place for your child to visit, especially when implementing some of the recommendations in this chapter. Check out 460 museums in fifty states and nineteen countries at https://finda childrensmuseum.org/, a site run by the Association of Children's Museums, to discover what might become your child's newest favorite destination. Many also offer activities for families to enjoy at home at this separate link: https://findachildrensmuseum.org/at-home/.

A FOCUS ON RAIL TRAVEL

In chapter 17, I've included a list of museums, divided by special interest, which may or may not have autism-friendly modifications in place. These include transit museums, but one compensating factor may be that these museums cater to one of the most popular interests of children with ASD, namely railroads and trains.

Theories abound as to the origins of this passion, whether it's to do with the predictability of the service, the uniformity of the tracks, or the spinning of the wheels. "I'm not sure I've ever heard anyone give a good explanation for how excited these children get about trains," says Tony Charman, professor of clinical child psychology at King's College London, who specializes in autism. "For different children it could be about the systemicity, the regular routine. But why trains, rather than cars? A car interest is much less common."[3] The topic is further discussed in chapter 10.

This intense interest has prompted venues like the New York Transit Museum to create a Subway Sleuths afterschool program that focuses on both the history of New York City transportation and putting children with autism at ease in social situations (https://www.nytransitmuseum.org/learn/subwaysleuths/). Similar train clubs have popped up—especially in England—that congregate and socialize together with other families with ASD at local transit museums. To find a list of over six hundred transit museums around the world to visit locally or while traveling, check out their Wikipedia listing at https://en.wikipedia.org/wiki/List_of_transport_museums.

AMUSEMENT PARKS (OTHER THAN DISNEY), WATER PARKS, AND SPORTING ARENAS

Many believe that amusement parks would be overwhelming for children on the spectrum. Betty J. Burger, recently retired from BGB Travel and Leisure, LLC, of St. Marys, Georgia, points to the large crowds as one possible factor. "Keep a child's sensory triggers in mind when considering various venues," she warns.

But surprisingly, many children with autism—those who normally eschew situations involving crowds and noise—have an interest in amusement parks and specifically roller coasters. So many so, in fact, that a growing number of theme parks now cater to this population.

One reason might be that trains are of particular interest to many on the spectrum and "roller coasters are sort of train-like," says Steven Crites, an associate professor of teacher education at Northern Kentucky University. "They run on tracks, they're repetitive, they make noise. All of those things are attractive to somebody with autism."[4]

The following are some strategies for making the theme park experience palatable:

Even though more and more amusement and theme parks are developing programs for those with autism, Nicole Thibault of Magic Storybook Travels in Fairport, New York, says you should still prepare the family ahead of time. "Explain where you are going and what you will see. Maybe show pictures of the park and their rides. You may also want to put together a bag with items that will help your child with any sensory issues. You could have little toys, hand sanitizer, scented lip balm. Anything that will help your child get through any unexpected issues," she suggests.[5]

That strategy proved successful for Jason Adlowitz, a parent from N. Chili, New York: "Early on in our Disney World adventures and even on a few trips to Universal Studios in Florida, we would watch movies on YouTube to get a [preview] of what the ride would be like. That way we were able to mitigate the risk ahead of time."

Once at the park, it's important to take safety measures, according to Kelli Engstrom, a parent from Kent, Washington. "We take a picture of our kids at the entrance. They think it's just a picture, but this is for us in case they get lost. We also pick a meet-up [location] and decide before we go in if we [will be visiting] the gift shop and [establish] how much will be spent."

Theme parks may exacerbate children's sensory issues, and it's important to take precautions ahead of time. Along with the sound levels that could require earplugs or noise-canceling headphones, parents should "be aware that many amusement rides feature flashing and spinning lights that may vary in intensity or kick up in intensity once the ride starts," warns Mark Hutten, the author of *My ASD Child*. "The concern here is that this constant 'strobe light' flickering may induce a meltdown in those kids who are overly sensitive to visual stimuli."[6] For that reason, along with checking for the days and times when attendance is at its lowest (most amusement parks keep track of this and will gladly share that information), Jonathan Haraty suggests that you make sure you bring along a pair of sunglasses for children on the spectrum, even if you are going to the park in the evening.

And even more than noise and visual stimulation, the waiting factor could try any child's patience, much less one who is neurodiverse. Says Michael Sokol: "Amusement parks are tough if lines are long. Some offer 'fast passes' or set up a reservation time to enter the ride. Do it! Sometimes concerts/shows will allow you to get in early before the masses arrive; all you need to do is ask. And at Hershey Park in Hershey, Pennsylvania, the Chocolate World Tour ride is free. One day we went on it twenty-nine times in a row!"

At theme parks, be sure to check in constantly with your youngster to gauge their reactions to their surroundings, says Mark Hutten. "At first, start slowly with gentle rides, paying careful attention for signs of over-stimulation. After each ride, process the experience with your youngster to gather his impressions and tolerance level. Some children absolutely relish the sensory feedback they derive from seemingly violent, whirling, spinning, upside-down-turning rides—but some don't, and after the ride is over, they may flip into a full-blown meltdown," he says.[7]

To head off that meltdown, Jonathan Haraty offers a tip for Bahamas-bound travelers:

> Water parks that are attached to resorts like Atlantis provide an excel-lent experience since they are capacity controlled. A hint for you: if you book a room at the Comfort Suites on Paradise Island, you get free admission to the Atlantis Resorts Waterpark. If you are on a cruise that docks in Nassau for most of the day, you can still reserve a room at that hotel. That way, you can spend the day at the waterpark for free, and if your child/young adult needs a break from the park, you are just a short walk to your room at the Comfort Suites. Think of it as a day room.

DISNEY

At Disney you can get a "stroller as wheelchair" sticker from Guest Services that will allow you to bring your stroller into queues where it wouldn't typically be allowed, says Michelle St. Pierre of Modern Travel Professionals in Yardley, Pennsylvania:

> No diagnosis is necessary; you just need to explain what it solves for your child. (Example: child can't stand still for long periods; stroller would allow her to sit while waiting. Or, child can't manage a crowded situation, stroller would allow him to sit and pull down the shade and have some isolation), I encourage families to look at the park or mu-seum maps ahead of time and suggest some quiet spots in case their kids need to decompress in the event they become overstimulated.

"Disneyland Resort has an attraction-by-attraction guide that will tell you which rides have loud noises, scents, and smells, flashing lights, etc. A good travel advisor can help you maximize the use of this and other tools," says Tara Woodbury.

Adrienne Query-Fiss, a parent from Seattle, Washington, offers accolades for Disney's customer service, which helped turn a potential disaster into a win:

> On our first trip to Disney World, we parked the car and walked up to the tram just as one was pulling away. My six-year-old, who is autistic and a catastrophizer, fell crying on the ground, thinking we'd never get to "hug Mickey Mouse." Like magic, a Cast Member in a yellow polo appeared out of thin air, sang him a song, did a dance, got him to stop crying, and then the tram pulled up. (All of this took place in like, ninety seconds max.) I thanked her profusely, loaded my kid, turned around to thank her one more time and poof—she was gone. Then the parking tram became his favorite part of Disney. We even bought him a model tram for his birthday that year.
>
> The Disney Access Pass, and Disney accommodations in general, are amazing. For non-Disney trips, we get a membership or multi-day pass so we can split in a hurry and not feel guilty. But Disney World is where it's at. Stay on property and treat it like a splurge. They're so accommodating.

Jason Adlowitz's family trips to Disney taught him many things, including making sure his son Bailey understands that ice cream is *not* available at nine a.m. in the park. His main Disney Tips for others include:

1. Get their DAS card★, which allows the disabled individual to wait anywhere else than stand in line. If the wait is sixty minutes, the return time would be less than that. Basically, the card freed us to find something else to do for fifty minutes rather than wait in line because I know Bailey becomes restless in lines.
2. Contact Hotel Guest Services ahead of time, provide your reservation number, and tell them you're traveling with someone with autism and would like a quiet room.
3. Be aware that Disney is like no other park, especially not Universal. Disney has Guest Services Cast Members all over the parks who are willing to engage and help with anything. We've gotten assistance with an array of items over the years and have never been told, "I don't know."
4. Always be in contact with your neurodiverse child. Check in with him or her regularly, especially regarding energy levels. Work with

your travel agent to fill out the dietary forms and make sure you get the [aforementioned] disability pass.

*At press time, Disney was making changes to its DAS program. It was unclear if an extra fee would be charged. As explained by Jennifer Hardy:

> Disney's biggest enhancement to the DAS program is that they will now allow you to "enroll" pre-arrival (up to 30 days in advance) via live video call rather than having to stand in line at Guest Services upon entry to the first theme park of your trip. During this call, you can also select your first two return times for each day of your visit. Also new to DAS, guests will be able to select attractions for return times directly via their mobile app rather than having to run across the theme park to request a return time from a Cast Member.

For more information on using DAS, including using it in conjunction with new programs like Disney Genie, Disney Genie+, and Lightning Lane, see https://disneyworld.disney.go.com/genie or contact Disney at https://disneyworld.disney.go.com/guest-services/disability-access-service/. You can also call them at (407) 560-2547.

All of the information shared in this chapter makes it clear that children on the spectrum can, with the right preparation and tools, enjoy the venues available to every neurotypical child. What follows are a list of autism-friendly and autism-certified venues that have already put in place programs to help you navigate any sensory or other issues; a discussion of the International Board of Credentialing and Continuing Education Standards (IBCCES) Accessibility Card (IAC), which can provide further assistance; and a tour company that organizes excursions specifically for families with children of all ability levels.

AUTISM-FRIENDLY VENUES AND CERTIFIED AUTISM CENTERS

The following are some theme parks, sports arenas, performance arts centers, water parks, museums, zoos, aquariums, and more in the United States—divided alphabetically by state and then by city—and in other countries. Those listed as "CAC" have received the IBCCES designation as Certified Autism Centers, while those listed as "AF" advertise autism-friendly programs for special needs visitors or have been designated by various organizations such as the Applied Behavior Analysis Programs Guide[8] as autism-friendly. Think of these as some but not all options as new venues announce their autism friendliness on an ongoing basis.

Full descriptions of CACs can be found in the most recent edition of the downloadable *Autism Adventure Guide* and *Autism Travel Directory* at AutismTravel.com. Changes were announced just as this book was going to press, so I've labeled any properties and venues that in 2022, for whatever reason, are no longer designated by IBCCES as Certified Autism Centers as "2021 CAC." Also note that venues marked as AF or 2021 CAC may vary widely in what they offer to the autistic population—for example, they may offer special ASD programming on days you won't be in town—so make sure to perform your due diligence before visiting. For the most current information, including additions to the list, be sure to couple the use of this book with visits to my blog, TravelingDifferent.com, and my Twitter feed at http://www.twitter.com/@Travelingdiff.

- Located in Multiple States
  - **Disney Parks and Resorts** (AF)
    https://disneyparks.disney.go.com

  - **Great Wolf Lodge** (AF)
    https://www.greatwolf.com

  - **LEGOLAND®** (AF, some CAC)
    https://www.legoland.com (Check with individual locations as to autism friendliness)

  - **SeaWorld** (AF)
    https://seaworld.com

  - **Six Flags Theme Parks** (CAC)
    https://www.sixflags.com/america

  - **Twenty-One Senses.org** (AF)
    A digital guide to venues with sensory friendly hours and accommodations
    https://www.twentyonesenses.org/sensory-friendly-hours

  - **Yellowstone National Park**—Montana, Idaho, Wyoming (AF)
    https://www.nps.gov/yell/index.htm

- Arizona:
  - **Arizona Goat Yoga** (2021 CAC)
    26601 S Val Vista Dr, Gilbert, AZ 85298
    480-269-4144
    https://goatyoga.com

  - **Oakland Athletics Spring Training Home** (CAC)
    Hohokam Stadium, 1235 N. Center Street, Mesa, AZ 85201
    480-644-7529
    https://www.mesaparks.com/parks-facilities/hohokam-stadium

  - **Mesa Arts Center** (CAC)
    1 E. Main St, Mesa, AZ 85201
    480-644-6500 (Box Office); MCA Museum 480-644-6560
    https://www.mesaartscenter.com

  - **Arizona Museum of Natural History** (CAC)
    53 N. Macdonald, Mesa, AZ 85201
    480-644-2230
    https://www.arizonamuseumofnaturalhistory.org

  - **I.D.E.A. Museum** (CAC)
    150 W. Pepper Pl, Mesa, AZ 85201
    480-644-2468
    https://www.ideamuseum.org

  - **All Aboard America! Bus Tours** (2021 CAC)
    230 S. Country Club Dr, Mesa, AZ 85210
    480-962-6202
    https://www.allaboardamerica.com

  - **Butterfly Wonderland** (2021 CAC)
    9500 East Vía de Ventura F100, Scottsdale, AZ 85256
    480-800-3000
    https://butterflywonderland.com

  - **Odysea Aquarium** (CAC)
    9500 East Vía de Ventura Suite A-100, Scottsdale, AZ 85256
    480-291-8000
    https://www.odyseaaquarium.com

- California
  - **Knott's Berry Farm** (AF)
    8039 Beach Boulevard, Buena Vista, CA 90620
    714-220-5200
    https://www.knotts.com

  - **The GRAMMY Museum L.A. Live** (CAC)
    800 W. Olympic Blvd, Los Angeles, CA 90015
    213-725-5700
    https://grammymuseum.org

  - **SeaWorld San Diego** (AF)
    500 Sea World Dr, San Diego, CA 92109
    619-222-4732
    https://seaworld.com/san-diego

  - **San Diego Zoo** (AF)
    2920 Zoo Dr, San Diego, CA 92101
    619-231-1515
    https://sandiegozoowildlifealliance.org

  - **Santa Barbara Zoo** (CAC)
    500 Ninos Dr, Santa Barbara, CA 93103
    805-962-5339
    https://www.sbzoo.org

  - **American Dream Parks** (AF)
    4701 Great America Pkwy, Santa Clara, CA 95054
    408-988-1776
    https://www.cagreatamerica.com

- Colorado
  - **Water World Colorado** (CAC)
    8801 N. Pecos Street, Federal Heights, CO 80260
    303-427-7873
    https://www.waterworldcolorado.com

  - **WOW! Children's Museum** (AF)
    110 N. Harrison Ave, Lafayette, CO 80026
    303-604-2424
    https://wowchildrensmuseum.org/

- **Great Sand Dunes National Park and Preserve** (AF)
  Visitor Center-11999 State Highway 150, Mosca, CO 81146
  719-378-6395
  https://www.nps.gov/grsa/index.htm

- District of Columbia
  - **Smithsonian Institution** (AF)
    The Smithsonian Institution is the world's largest museum, education, and research complex, with nineteen museums and the National Zoo. These include the National Air and Space Museum and the National Museum of African American History and Culture. Addresses vary.
    202-633-1000
    https://www.si.edu/museums

- Florida
  - **Fort Myers Mighty Mussels—Hammond Stadium** (2021 CAC)
    14400 Six Mile Cypress Pkwy, Fort Myers, FL 33912
    239-768-4210
    https://www.milb.com/fort-myers/ballpark/hammondstadium

  - **Museum of Contemporary Art** (AF)
    333 N. Laura St, Jacksonville, FL 32202
    904-366-6911
    https://mocajacksonville.unf.edu

  - **Island H2O Live! Water Park** (CAC)
    3230 Inspiration Dr, Kissimmee, FL 34747
    407-910-1401
    https://www.sunsetwalk.com/island-h2o-live-water-park

  - **Alligator and Wildlife Discovery Center** (2021 CAC)
    12973 Village Blvd, Suites A-D, John's Pass, Madeira Beach, FL 33708
    727-329-8751
    https://kissagator.com

  - **Miami Children's Museum** (AF)
    980 MacArthur Causeway, Miami, FL 33132
    305-373-5437
    https://www.miamichildrensmuseum.org/

- ○ **The Golisano Children's Museum** (AF)
  15080 Livingston Rd, Naples, FL 34109
  239-514-0084
  https://cmon.org

- ○ **Dr. Phillips Center for the Performing Arts** (CAC)
  445 S. Magnolia Ave, Orlando, FL 32801
  407-839-0119
  https://www.drphillipscenter.org

- ○ **Aquatica Orlando Water Park** (CAC)
  5800 Water Play Way, Orlando, FL 32821
  407-545-5550
  https://aquatica.com/orlando/

- ○ **Discovery Cove Theme Park** (CAC)
  6000 Discovery Cove Way, Orlando, FL 32821
  407-513-4600
  https://discoverycove.com/orlando

- ○ **SeaWorld Orlando** (2021 CAC)
  7007 Sea World Dr, Orlando, FL 32821
  407-545-5550
  https://seaworld.com/orlando

- ○ **Glazer Museum** (AF)
  110 W. Gasparilla Plaza, Tampa, FL 33602
  813-443-3861
  https://glazermuseum.org

- • Georgia
  - ○ **Georgia Aquarium** (2021 CAC)
    225 Baker St NW, Atlanta, GA 30313
    404-581-4000
    https://www.georgiaaquarium.org

  - ○ **Chick-fil-A College Football Hall of Fame** (AF)
    250 Marietta St NW, Atlanta, GA 30313
    404-880-4800
    https://www.cfbhall.com

- ○ **The Children's Museum of Atlanta** (AF)
  275 Centennial Olympic Park Dr NW, Atlanta, GA 30313
  404-659-5437
  https://childrensmuseumatlanta.org

- Illinois
  - ○ **Four Rivers Environmental Education Center** (CAC)
    25055 W. Walnut Ln, Channahon, IL 60410
    815-722-9470
    https://www.reconnectwithnature.org/preserves-trails/visitor
    -centers/four-rivers-environmental-education-center

  - ○ **Shedd Aquarium** (AF)
    1200 S. Lake Shore Dr, Chicago, IL 60605
    312-939-2438
    https://www.sheddaquarium.org

  - ○ **Chicago Children's Museum at Navy Pier** (AF)
    700 E. Grand Ave, Suite 127, Chicago, IL 60611
    312-527-1000
    https://www.chicagochildrensmuseum.org

  - ○ **DuPage Children's Museum** (AF)
    301 N. Washington St, Naperville, IL 60540
    630-637-8000
    https://dupagechildrens.org

- Indiana
  - ○ **Children's Museum of Indianapolis** (AF)
    3000 N. Meridian St, Indianapolis, IN 46208
    317-334-4000
    https://www.childrensmuseum.org

  - ○ **Holiday World and Splashin' Safari Park** (AF)
    452 E. Christmas Blvd, Santa Claus, IN 47579
    800-467-2682
    https://www.holidayworld.com

- Iowa
  - **Family Museum** (CAC)
    2900 Learning Campus Dr, Bettendorf, IA 52772
    563-344-4106
    https://www.familymuseum.org

- Maryland
  - **The Walters Art Museum** (AF)
    600 N. Charles St, Baltimore, MD 21201
    410-547-9000
    https://thewalters.org

  - **Spinners Pinball Arcade: A Pinball EDU Center** (CAC)
    919 N. East Market Street, Frederick, MD 21701
    No Phone Number Available
    https://www.spinnerspinball.org

- Massachusetts
  - **Edaville Family Theme Park** (AF)
    5 Pine St, Carver, MA 02330
    508-866-8190
    http://www.edaville.com

  - **Boston Children's Museum** (AF)
    308 Congress St, Boston, MA 02210
    617-426-6500
    https://www.bostonchildrensmuseum.org

  - **Museum of Fine Arts** (AF)
    465 Huntington Ave, Boston, MA 02115
    617-267-9300
    https://www.mfa.org

- Michigan
  - **Soaring Eagle Waterpark and Hotel** (CAC)
    5665 E. Pickard Rd, Mt. Pleasant, MI 48858
    989-817-4800
    https://www.soaringeaglewaterpark.com

- ° **Michigan's Adventure Theme Park** (AF)
  1198 W. Riley-Thompson Rd, Muskegon, MI 49445
  231-766-3377
  https://www.miadventure.com

- ° **Bowlero Lanes and Lounge** (CAC)
  4209 Coolidge Hwy, Royal Oak, MI 48073
  248-549-7500
  https://bowlerodetroit.com

- Minnesota
  - ° **Mall of America** (CAC)
    2131 Lindau Lane, Bloomington, MN 55425
    952-883-8800
    https://mallofamerica.com/
    (Many attractions, including the Crayola Experience and Nickelodeon Universe are listed separately below.)

  - ° **Crayola Experience** (2021 CAC)
    300 South Avenue, Level 3, Bloomington, MN 55425
    952-851-5800
    https://www.mallofamerica.com/directory/crayola-experience

  - ° **Nickelodeon Universe** (CAC)
    5000 Center Ct, Bloomington, MN 55425
    952-883-8555
    https://nickelodeonuniverse.com

  - ° **ValleyFair Amusement Park** (2021 CAC)
    1 Valley Fair Dr, Shakopee, MN 55379
    952-445-7600
    https://www.valleyfair.com

- Missouri
  - ° **Ripley's Believe It or Not! and Super Fun Zone of Branson** (2021 CAC)
    3326 76 Country Blvd, Branson, MO 65616
    417-337-5300
    https://www.ripleys.com/branson

- **Dogwood Canyon Nature Park** (2021 CAC)
  2038 West State Hwy 86, Lampe, MO 65681
  877-459-5687
  https://dogwoodcanyon.org

- **Worlds of Fun** (AF)
  4545 Worlds of Fun Avenue, Kansas City, MO 64161
  816-454-4545
  https://www.worldsoffun.com

- **Johnny Morris' Wonders of Wildlife National Museum & Aquarium** (2021 CAC)
  500 W. Sunshine St, Springfield, MO 65807
  888-222-6060
  https://wondersofwildlife.org

- New Hampshire
  - **Story Land Theme Park** (2021 CAC)
    850 NH Route 16, Glen, NH 03838
    603-383-4186
    https://www.storylandnh.com

  - **Children's Museum of New Hampshire** (AF)
    6 Washington St, Dover, NH 03820
    603-742-2002
    https://childrens-museum.org

  - **Cheshire Children's Museum** (AF)
    149 Emerald St, Keene, NH 03431
    603-903-1800
    https://www.cheshirechildrensmuseum.org

- New York
  - **Explore & More, The Ralph C. Wilson, Jr. Children's Museum** (AF)
    130 Main St, Buffalo, NY 14202
    716-655-5131
    https://exploreandmore.org

- **Splish Splash Water Park** (CAC)
  2549 Splish Splash Dr, Calverton, NY 11933
  631-727-3600
  https://www.splishsplash.com

- **Six Flags Darien Lake** (AF)
  9993 Allegheny Rd, Corfu, NY 14036
  585-599-4641
  https://www.sixflags.com/darienlake

- **The Friends Experience** (CAC)
  130 E 23rd St, New York, NY 10010
  No phone number available, contact form on website
  https://www.friendstheexperience.com/new-york

- **Intrepid Sea, Air & Space Museum** (AF)
  Pier 86, W. 46th St, New York, NY 10036
  212-245-0072
  https://www.intrepidmuseum.org

- **American Museum of Natural History** (AF)
  200 Central Park West, New York, NY 10024
  212-769-5100
  https://www.amnh.org

- **New York Transit Museum** (AF)
  99 Schermerhorn St, Brooklyn, NY 11201
  718-694-1600
  https://www.nytransitmuseum.org

- **Museum of Modern Art** (AF)
  11 W. 53rd St, New York, NY 10019
  212-708-9400
  https://www.moma.org

- **The Metropolitan Museum of Art** (AF)
  1000 Fifth Ave, New York, NY 10028
  212-535-7710
  https://www.metmuseum.org

- **Children's Museum of the Arts** (AF)
  103 Charlton St, New York, NY 10014
  212-274-0986
  https://cmany.org

- **Children's Museum of Manhattan** (AF)
  The Tisch Building, 212 W 83rd St, New York, NY 10024
  212-721-1223
  https://cmom.org

- **The Jewish Museum** (AF)
  1109 Fifth Ave, New York, NY 10128
  212-423-3200
  https://www.thejewishmuseum.org

- **Whitney Museum of American Art** (AF)
  99 Gansevoort St, New York, NY 10014
  212-570-3600
  https://whitney.org

- **Central Park Zoo/Wildlife Conservation Society** (AF)
  Fifth Avenue and 64th Street, New York, NY 10065
  212-439-6500
  https://centralparkzoo.com

- **Solomon R. Guggenheim Museum** (AF)
  1071 Fifth Ave, New York, NY 10128
  212-423-3500
  https://www.guggenheim.org

- **Brooklyn Children's Museum** (AF)
  145 Brooklyn Ave, Brooklyn, NY 11213
  718-735-4400
  https://www.brooklynkids.org

- **The Strong National Museum of Play** (AF)
  One Manhattan Square Dr, Rochester, NY 14607
  585-263-2700
  https://www.museumofplay.org

- North Carolina
  - **Carowinds Theme Park** (AF)
    14523 Carowinds Blvd. Charlotte, NC 28273
    704-588-2600
    https://www.carowinds.com

  - **North Carolina Maritime Museum at Southport** (CAC)
    204 E. Moore St, Southport, NC 28461
    910-477-5150
    https://ncmaritimemuseumsouthport.com

- Ohio
  - **Kings Island Amusement Park** (2021 CAC)
    6300 Kings Island Dr, Kings Island, OH 45034
    513-754-5700
    https://www.visitkingsisland.com

  - **Cedar Point Amusement Park** (AF)
    1 Cedar Point Dr, Sandusky, OH 44870
    419-627-2350
    https://www.cedarpoint.com

  - **Butler Institute of American Art** (AF)
    524 Wick Ave, Youngstown, OH 44502
    330-743-1107
    https://butlerart.com

- Oklahoma
  - **Frontier City** (CAC)
    11501 N 1-35 Service Road, Oklahoma City, OK 73131
    405-478-2140
    https://www.sixflags.com/frontiercity

- Oregon
  - **Jordan Schnitzer Museum of Art** (AF)
    University of Oregon, 1430 Johnson Lane, Eugene, OR 97403
    541-346-3027
    https://jsma.uoregon.edu

- Pennsylvania
  - **Dorney Park and Whitewater Kingdom** (2021 CAC)
    4000 Dorney Park Rd, Allentown, PA 18104
    610-395-3724
    https://www.dorneypark.com

  - **Turkey Hill Experience** (CAC)
    301 Linden St, Columbia, PA 17512
    844-847-4884
    https://www.turkeyhillexperience.com

  - **Knoebels Amusement Resort** (CAC)
    391 Knoebels Boulevard, PA-487, Elysburg, PA 17824
    570-672-2572
    https://www.knoebels.com

  - **Hershey's Chocolate World Attraction** (CAC)
    101 Chocolate World Way, Hershey, PA 17033
    717-534-4900
    https://www.chocolateworld.com

  - **Sesame Place Philadelphia** (CAC)
    100 Sesame Rd, Langhorne, PA 19047
    215-702-3566
    https://sesameplace.com/philadelphia

  - **Elmwood Park Zoo** (CAC)
    1661 Harding Blvd, Norristown, PA 19401
    800-652-4143
    https://www.elmwoodparkzoo.org

  - **Museum of the American Revolution** (CAC)
    101 S. 3rd St, Philadelphia, PA 19106
    215-253-6731
    https://www.amrevmuseum.org

  - **Please Touch Museum** (AF)
    Memorial Hall, 4231 Avenue of the Republic, Philadelphia, PA 19131
    215-581-3181
    https://www.pleasetouchmuseum.org

- Andy Warhol Museum (AF)
  117 Sandusky St, Pittsburgh, PA 15212
  412-237-8300
  https://www.warhol.org

- LEGOLAND Discovery Center Philadelphia (an indoor attraction) (AF)
  500 W Germantown Pike Unit #1055, Plymouth Meeting, PA 19462
  267-245-9696
  https://www.legolanddiscoverycenter.com/philadelphia

- Children's Museum of Pittsburgh (AF)
  10 Children's Way, Allegheny Square, Pittsburgh, PA 15212
  412-322-5058
  https://pittsburghkids.org

- Reading Public Museum (CAC)
  500 Museum Rd, Reading, PA 19611
  610-371-5850
  https://www.readingpublicmuseum.org

- Kennywood Theme and Amusement Park (2021 CAC)
  4800 Kennywood Blvd, West Mifflin, PA 15122
  412-461-0500
  https://www.kennywood.com

- South Carolina
  - Ripley's Aquarium of Myrtle Beach (2021 CAC)
    1110 Celebrity Circle, Myrtle Beach, SC 29577
    843-916-0888
    https://www.ripleyaquariums.com/myrtlebeach

- Tennessee
  - Creative Discovery Museum (AF)
    321 Chestnut St, Chattanooga, TN 37402
    423-756-2738
    https://www.cdmfun.org

- ○ **Ripley's Aquarium of the Smokies** (2021 CAC)
  88 River Rd, Gatlinburg, TN 37738
  865-430-8808
  https://www.ripleyaquariums.com/gatlinburg

- ○ **Dollywood** (AF)
  2700 Dollywood Parks Blvd, Pigeon Forge, TN 37863
  800-365-5996
  https://www.dollywood.com

- Texas
  - ○ **Dallas Museum of Art** (AF)
    1717 N. Harwood St, Dallas, TX 75201
    214-922-1200
    https://dma.org

  - ○ **Fort Worth Zoo** (CAC)
    1989 Colonial Pkwy, Fort Worth, TX 76110
    817-759-7555
    https://www.fortworthzoo.org

  - ○ **Schlitterbahn Galveston** (AF)
    2109 Gene Lucas Blvd, Galveston, TX 77554
    409-770-9283
    https://www.schlitterbahn.com

  - ○ **Children's Museum of Houston** (AF)
    1500 Binz St, Houston, TX 77004
    713-522-1138
    https://www.cmhouston.org

  - ○ **Houston Museum of Natural Science** (CAC)
    5555 Hermann Park Dr, Houston, TX 77030
    713-639-4629
    http://www.hmns.org

  - ○ **Space Center Houston** (CAC)
    1601 E. NASA Pkwy, Houston, TX 77058
    281-283-4755
    https://spacecenter.org

- ○ **Schlitterbahn New Braunfels** (AF)
  381 E. Austin St, New Braunfels, TX 78130
  830-625-2351
  https://www.schlitterbahn.com

- ○ **Morgan's Wonderland Theme Park** (AF)
  5223 David Edwards Dr, San Antonio, TX 78233
  210-495-5888
  https://morganswonderland.com

- Utah
  - ○ **National Ability Center** (AF)
    1000 Ability Way, Park City, UT 84060
    435-649-3991
    https://discovernac.org/programs

- Virginia
  - ○ **Kings Dominion and Soak City** (AF)
    16000 Theme Park Way, Doswell, VA 23047
    804-876-5000
    https://www.kingsdominion.com

  - ○ **Wonder Universe, A Children's Museum** (2021 CAC)
    782 New River Rd #812, Christiansburg, VA 24073
    540-200-8231
    https://wonderuniverse.org

  - ○ **Mount Vernon** (AF)
    3200 Mount Vernon Memorial Hwy, Mount Vernon, VA 22121
    703-780-2000
    https://www.mountvernon.org

- Washington
  - ○ **KidsQuest Children's Museum** (AF)
    1116 108th Ave NE, Bellevue, WA 98004
    425-637-8100
    https://www.kidsquestmuseum.org

  - ○ **Hands On Children's Museum** (AF)
    414 Jefferson St NE, Olympia, WA 98501
    360-956-0818
    https://www.hocm.org

- **Pacific Science Center** (AF)
  200 2nd Ave N, Seattle, WA 98109
  206-443-2001
  https://www.pacificsciencecenter.org

- **Seattle Children's Museum** (AF)
  305 Harrison St, Seattle, WA 98109
  206-441-1768
  https://thechildrensmuseum.org

- Wisconsin
  - **Wilderness Resort Hotel and Waterparks** (AF)
    511 E. Adams, Wisconsin Dells, WI 53965
    800-867-9453
    https://www.wildernessresort.com

- Outside the United States
  - Canada
    - **MarineLand Canada** (AF)
      7657 Portage Rd, Niagara Falls, ON L2E 6X8, Canada
      905-356-9565
      https://www.marineland.ca

    - **Ripley's Aquarium of Canada** (CAC)
      288 Bremner Blvd, Toronto, ON M5V 3L9, Canada
      647-351-3474
      https://www.ripleyaquariums.com/canada

    - **Six Flags La Ronde** (CAC)
      22 Chemin Macdonald, Montreal, Quebec, H3C 6A3 Canada
      514-397-2000
      https://www.sixflags.com/larondeen

  - Lithuania
    - **UNO Parks Vilnius** (CAC)
      Lizdeikos str. Antakainis, Vilnius 10101, Lithuania
      370-602-32366
      http://www.unoparks.lt

    ◦ Qatar
- **KidzMondo Doha** (CAC)
  Mall of Qatar, Dukhan Highway, Building 800 Street 373
  Doha, Qatar
  974-4028-5888
  https://www.facebook.com/KidzMondoDoha

AN IAC PRIMER

As more and more hotels, resorts, and attractions become autism certified, the IBCCES has developed a method to increase satisfaction and reduce potential abuse of benefits. The IAC is a free identification card that can be requested online. It is "designed to help individuals with cognitive disorders or physical impairments receive helpful accommodations at certified attractions worldwide."[9] Those who are eligible for the card include those who suffer from Alzheimer's, autism, dementia, Down syndrome, post-traumatic stress disorder, bipolar disorder, attention-deficit/hyperactivity disorder, or those who have limited mobility or hearing or visual impairments. Once an applicant is approved, the IAC card can be downloaded digitally but there is also an option to print it, if preferred. It is good for one year and can be updated or renewed as often as necessary.

While this can make families on the spectrum aware of autism-friendly locales and reduce delays and confusion when checking in, IBCCES makes it clear that the IAC is not a ticket. It does not guarantee entry to an attraction or any special accommodations or benefits; those are provided at the sole discretion of the venue, whose Guest Services desk will describe those accommodations at the time the card is presented.

To obtain a card, you'll need a recent photo of the cardholder, general information about the cardholder's cognitive disorders or disabilities, and any accommodations requested. Applicants must also provide contact information for the parent, guardian, or caretaker of the cardholder, as well as contact information for their health care provider (doctor, therapist, or counselor). That provider must provide a statement regarding the extent and circumstances of the disorder.

ORGANIZED TOURS FOR
FAMILIES ON THE SPECTRUM

Most of the advice in this book pertains to individual families traveling on their own, except in the case of cruises like the ones organized by Autism on the Seas (more in chapter 11). But what about families who might want to meet others navigating similar issues and enjoy the companionship of traveling together? What resources are available to them?

Unfortunately, very few tour operators specialize in group tours for special needs families. Those that do are usually travel agencies that offer curated group travel with specific itineraries that are geared toward those with special needs. Lisa Bertuccio of Vacation Your World in Otisville, New York, reports that she's in the initial phase of putting together one of these group tours and hopes to offer more of them in the future.

The logistics aren't that different from organizing any group itinerary, she says, but such tours require more attention to detail, such as securing wheelchairs or assisted devices in advance, having a daily specific itinerary, notifying the venue or park in advance to secure reserved seating etc., and assigning a group leader who's skilled in working with this population.

"The main benefit over individual travel would be the social aspect," says Bertuccio:

> Often, it's easier to teach or [demonstrate] certain skills or behaviors when you're in a group setting. It's also wonderful for parents or caregivers to have the support of others within the group should a hiccup occur. You understand what the other parent is going through if one of the kids has a meltdown. If the parent isn't sure of how to handle a behavior, another parent may have had a similar experience and can help.
>
> Other benefits would be that group rates are often cheaper than if you traveled on your own, there is usually reserved seating for groups, and things are much more [organized] ahead of time, so parents, caregivers, children, and other participants know what to expect on their vacation. There are also group departments at theme parks, hotels, and other venues that aid the travel agent in customizing the vacation for the group. Having a point person to cure problems that arise is really key.

One tour operator that *does* organize group tours for those with special needs is Wilderness Inquiry. Says Nell Holden, their business operations director: "Wilderness Inquiry was founded in 1978 [focusing on] integrated wilderness trips for all travelers. Their stated mission is 'to connect people of all ages, backgrounds, and abilities to each other and the natural world through shared outdoor adventures.' . . . Our adventures encourage people to open themselves to new possibilities and opportunities."[10]

From the beginning, Wilderness Inquiry has welcomed and supported all individuals, including those with autism and other social or emotional differences. They created their Families Integrating Together program in 1993, with funding from the U.S. Department of Education, to assist families with a parent or child with a disability. Their rationale: such families may not have previously been able to travel or take vacations together because they often "lack the means, know-how, or energy to plan and conduct multi-day family outings, [or] enjoy such a trip [without] the additional support [Wilderness Inquiry can provide] in the form of staff, personal care attendants, adapted equipment, and financial assistance to help make the outdoors accessible [to everyone]."[11]

"Our family trips are open for all families," says Holden:

> Frequently at least one member of each family on a trip has a disability but not always. Our goal is social integration and we have found that people are frequently most pleased with their connections to each other beyond just the outdoor experience. In addition, we have a program called Gateway to Adventure which serves individuals with intellectual and developmental disabilities who need additional practice or support in outdoor environments.
>
> Every itinerary is different, and every person is different. Most of our trips are designed so that everyone can participate. We can frequently offer additional support or adaptive equipment to make this possible. We have a few itineraries that may be more challenging for certain individuals; however even with these trips, we can often [implement] certain accommodations to make it work.

The challenge? "Often what gets in [someone's] way is [their] fear or hesitation. Being open to new experiences and willing to give it a try is a huge first step. Parents are frequently very surprised at what their children are capable of while on a Wilderness Inquiry trip."

The benefit?

We take each individual as they come and really cater our assistance to the specific needs of the individual and the group. Our staff goes through in-depth training to learn how to work with diverse populations. We have adaptive equipment and can be flexible in our planning and itineraries to accommodate all levels. Through social integration and human connectedness during a wilderness trip, the biggest growth and learning opportunities usually [revolve] around social and emotional well-being. It can be a powerful experience to accomplish a goal as a team with people of all abilities. When we are living and traveling outdoors together, we discover our common humanity.

Wilderness Inquiry's tours for families range from one to six days and include tours to the Mississippi River, St. Croix, Yellowstone, and Glacier National Park. Holden adds:

A very popular destination is the Apostle Islands [in Wisconsin]. We have a base camp in the area, and this allows families a few extra creature comforts while they are camping. There are showers, refrigeration, and platform tents. It makes the experience much more accessible than some areas that are more remote. Yellowstone is also very popular because there's so much to do and see in the park. The itinerary can be adjusted to fit the needs and desires of the group.

Wilderness Inquiry family trips usually accommodate four to eighteen participants with two to three staff members. Ratios can differ based on the itinerary, means of travel, and the needs of the group. For more information, visit their website at https://www.wildernessinquiry.org/ or call 612-676-9400.

PARENT SPOTLIGHT—
SARAH LUDWIG, NYACK, NEW YORK

Sarah Ludwig is the mother of two children, one on the spectrum diagnosed with attention-deficit/hyperactivity disorder and ASD. She's been traveling with the oldest since he was an infant, either for leisure or to see relatives, though she had concerns before starting: "Anytime we are outside of routine, he has challenges. I worry about public meltdowns and inconveniencing other travelers, but mostly his comfort and well-being."

"I think the greatest misconceptions are not limited to travel. People do not necessarily understand what autism looks like and that there are a wide range of experiences. My child is verbal and does well masking socially, for example, but he still struggles."

Traveling with one neurotypical and one neurodiverse child has its challenges, says Ludwig:

> My neurotypical daughter is exceptionally easygoing; she is flexible and small challenges don't bother her at all. I have to remember to slow down and be calm for my son and help him with issues that I may not have anticipated would bother him. I think being super-organized generally helps everyone feel confident.
>
> Just like any other family, we keep our children's best interests in mind when choosing a destination. Camping, at this point for example, would just make my son absolutely miserable. And while we would love to travel globally, the travel itself is still extremely hard and draining on Cole and adjusting to a whole foreign culture and time zone would be overwhelming. We google our destination [followed by the word] "autism" to see if there are services or programs specifically tailored to assisting. We like to stay at Airbnbs or rent condos so there is a kitchen, and we can eat in at least some of the time. We show pictures of the rental to Cole ahead of time, so he knows what to expect. Going out for meals every time is too much, but if we do dine out, we try to preview the menu and go at a quieter time.

Her advice to other parents traveling with children on the spectrum: "Give yourself some grace. Sometimes it feels like the world is watching you, but most people are too wrapped up in themselves to be focused on others and even if your child *is* having an attention-grabbing meltdown, it will pass."

CHAPTER 14 ACTION STEPS

1. When scheduling any tour or visit to a theme park, museum, or venue, keep your child's attention and energy levels in mind. You may not be able to do everything all at once, and that's okay. Incorporate planned breaks into your agenda.
2. Seek out venues that are autism-friendly or autism-certified. They may have lowered lighting, fewer people, less noise, and sensory

rooms or quiet family rooms where you can retreat and recharge. Do your research before leaving home.

3. Ask what days/times are least crowded. Many venues keep track and are happy to share this information. You can also ask which attractions or exhibits should be avoided by those with your child's sensory profile.

4. Choose the vacation that's best for your child. Disney is not a rite of passage. If you are beach people, don't head to the city and vice versa.

5. Use Social Stories to prepare the child for the adventure.

6. Preview new sensations before leaving home. If your child is unfamiliar with the feel of sand, buy some and experiment with it before booking a beach vacation.

7. Always check to see if your venue has special provisions for neurodiverse children. This includes varied food options, special access lines, the ability to use your stroller-as-wheelchair, sensory guides, sensitivity trained staff, readmittance passes (if overstimulation makes you leave early), and more. The IAC can also help.

8. Along with the usual, like fidget toys, headphones/earplugs, and snacks, make sure to have sunglasses on hand, even at night, to tamp down the visual stimulation at theme parks.

9. To avoid meltdowns caused by overstimulation, constantly check in with your child.

10. Consider hiring a nanny for the day for support and an extra set of hands.

11. Take the usual safety precautions, such as writing a caregiver's contact information on the child's arm or placing a GPS tracker in their clothes. Take photos at the park entrance so you will have a picture to circulate (with the child wearing his or her current clothing) should you become separated.

12. On sightseeing tours, provide children with their own headsets to hear the narrative or hire a personal guide. Choose tours that offer priority access. Let the tour guide know your child's particular interests so he or she can steer the narrative in that direction. Or skip the tours and travel to attractions independently.

13. If you think an organized tour might be a great way to meet other families with similar challenges, check out Wilderness Inquiry.

15

TAKING THE BITE OUT
OF RESTAURANT DINING

"When Xenia was small, if restaurants were too dark, cold, hot, or smelly, she would throw the biggest tantrums to the point where I would have to get up and immediately leave. Then when I got her official diagnosis at 2.5 years old, I knew what was going on. So, I observed her triggers, i.e., refrigerators buzzing, loud music, etc. and I would not take her to restaurants like those. As I became more knowledgeable, I would slowly introduce more conducive places and pick brunch hours so she would get the feel of the restaurant without high capacity. Then I would bring her during dinner with a comfort item but I would still request a booth away from the main dining room and close to a bathroom so she could have a quiet place if she got overwhelmed." – Delight Dolcine, parent, Brooklyn, New York

For many families on the spectrum, visiting restaurants while on vacation can be a disaster waiting to happen. Consider crowds, noise, unfamiliar smells, and foods that differ from your child's favorites, especially if you're traveling abroad. Not to mention having your child confined to a highchair and eliciting nasty looks from neighboring tables when tantrums erupt. It's as far from a relaxing repast as one can imagine, and if you haven't rented accommodations with a kitchen, you may be condemned to repeat it three times a day for the duration of your stay.

"Avoiding dining out while on vacation can be tough," says Tara Woodbury of Escape into Travel in Leland, North Carolina. "[But] now more than ever, restaurants are becoming better at providing not only menus that meet the needs of patrons with autism and other special needs, but also training to their staff, and services to make meals more enjoyable for families with someone on the spectrum."

"Highchairs were sometimes difficult to come by when Thessaly was younger," recounts Michelle Zeihr, a parent from Barrie, Ontario, Canada:

> If the restaurant did have them, some didn't have buckles, which allowed her to stand up or try to escape. If you can bring your own food, great. Most places were kind enough to allow us to bring Thessaly's meals in while we ate ours. They would often heat the food up for us if needed. Also, don't be afraid to make something your own from the menu if you see they have the ingredients. For instance, at the buffet you see someone making toast. The other end of the buffet has cheese. You can gather these things and bring them to the cook and ask if they will make a grilled cheese sandwich for your child. . . . Luckily, everywhere we've traveled has had pizza available for delivery!

The most recent edition of the *Autism Adventure Guide*, downloadable from the Autism Travel site (https://autismtravel.com/autismadventure guide-signup-2021) lists a limited number of restaurants bearing the Certified Autism Center designation. In addition, the "Autism Village" app carries reviews by patrons of restaurants and their suitability for those on the spectrum. The app is available free of charge at http://about.autismvillage.com/app.

One example of a family restaurant catering to the autistic community is Riv's Toms River Hub in Ocean County, New Jersey. Riv's has a separate dining room with a private entrance called "Chase's Friends Zone" that can be reserved with twenty-four hours' notice. Seating up to forty-five diners, it features specially trained waiters, soothing lighting, a corner with beanbag chairs, lots of sensory activities to keep children occupied, and a digital menu with pictures of the dishes, including gluten-free entrees and dairy-free cakes.[1]

Just because a restaurant may not be as autism friendly as Riv's, it doesn't mean meals eaten outside the home have to be stressful. Here are some tips from parents and experts on how to survive and even thrive during restaurant visits:

PREVIEW THE EXPERIENCE

"I remember back when I worked in classrooms teaching social skills," says Jodi Heckman-Caliciotti, special education teacher and behavior specialist, Rockland County, New York:

I knew one of my students was going to a restaurant for the first time. I showed pictures of restaurants and described the roles of the greeters, servers, chefs, cashiers, and the diners. Then we set the room up like a restaurant and took turns acting out different roles. We played restaurant sounds and practiced waiting to be seated, ordering off a menu, and speaking at an acceptable volume. All the kids loved it so much and my student did awesome on his restaurant visit!

You can do the same thing at home, according to Angela Romans of Rejuvia Travel in South Bend, Indiana. "Practice a dry run of eating out. Review the process of looking over menus, ordering food, sitting, and coloring at the table. Reinforce the idea that it is important to stay seated. Attempt to have your child use the bathroom at home beforehand to avoid [toileting] issues at the restaurant," she says.

START SMALL AND ALERT THE STAFF

One way to dip a toe in is to sample restaurants near your home before attempting to eat out while traveling. Perhaps visit a local eatery during their slowest period (between lunch and dinner might be a perfect time) and only stay for appetizers or dessert. You can always build from there.

Tara Woodbury recommends calling in advance and explaining to the restaurant staff why you would like to be seated quickly and why your neurodiverse traveler might not be able to answer the waitstaff's questions. But if you can't call beforehand, Carl W. Johnson of Adventures in Golf (incorporated as Hamill Travel Services, Inc.) in Amherst, New Hampshire, recommends speaking with the server as you enter and explaining that you need a quiet corner, or even something simple like the person's favorite drink or snack (e.g., French fries) served upon arrival. "That way, the child on the spectrum will almost immediately realize it could be a fun outing—favorite drink and/or snack—not a bad thing!"

However, he warns that not everything goes as planned and families should be prepared for that. "We've had to leave restaurants in the past—they were simply too loud, or the lighting was too irritating. You just get up, let the server know the situation, then go. Show the child that their inabilities to cope with a stressful environment were heard loud and clear, with the stressful situation removed (by you leaving the restaurant)," he says.

PLANNING IS THE KEY

"Kid-friendly, kid-friendly, kid-friendly," says Nicole Graham Wallace of Trendy Travels by Nicole—Dream Vacations in McDonough, Georgia. "Leave the fancy restaurants behind because children—on the spectrum or not—won't stay quiet and calm for long, especially in dress-up clothes."

Plan your meals out far enough in advance so that you aren't deviating from a schedule that your child is used to, suggests Michelle St. Pierre of Modern Travel Professionals in Yardley, Pennsylvania. "This is a good tip for *all* kids. People don't realize when they travel that sometimes planning can make a world of difference in everyone's temperament," she says.

This planning is made easier by the fact that many restaurants will post their entire menu on their websites. "Reviewing them before you visit allows you to eliminate specific dishes or select your meals in advance," says Jennifer Hardy of Cruise Planners in Kent, Washington:

> It will help you order faster and reduces the amount of time you wait for your food. It also sets the expectations of what is and is not available for your child to order in advance. The worst thing you could do is sit down at a restaurant with a hungry child who only eats mac and cheese, only to find out that [it's not offered] at that location.
>
> While many family-friendly chains may offer this automatically, let your server know if you need the child's meals to come out faster, usually with the appetizer course, rather than with the adult meals. This can help prevent a meltdown while waiting. Alternatively, consider bringing small snacks as allowed by the restaurant.

HANDLING THE SENSORY INPUT AND IMPATIENCE

"Restaurants can be extremely loud and busy for our son, Nate," says Michael Sokol of Longmeadow, Massachusetts. "We usually opt for an early dinner at four or four-thirty. If this is not possible, we bring an iPad, DS, iPhone, some device to keep him entertained and somewhat distracted from the environment. I will often ask to be seated next to a wall or a window, and never in the middle of the dining area (too loud and overwhelming). As a last resort, there's always takeout."

The opposite is true for Belinda S. Stull of Travel Leaders/All About Travel, LLC, in Hagerstown, Maryland, who says in her family, they look for a restaurant that has a somewhat noisy atmosphere since their grandson can be loud and excitable, which can be drowned out by the din around them. "However, we have also dined with him at quieter locations and have redirected him at times," she relates.

"Another tip that works for all children is to ask to be seated in a corner booth if possible. These seats are very roomy and set back from the traffic of customers walking to and from their seats," says Lisa Bertuccio of Vacation Your World in Otisville, New York.

Many people on the spectrum cannot handle dozens of conversations happening at the same time, so larger, buffet-style restaurants can be more problematic than a small restaurant with only five or ten tables, advises Mark Hutten, author of *My ASD Child*:

> Look for restaurants with patios so you can sit outside. It's usually not as crowded, and there is a little more space between tables. Give the server your credit card upfront and tell him or her you may have to leave early. Watch for signs of an impending meltdown (e.g., child is holding his head, frowning, getting fidgety, beginning to exhibit "tics" such as rapid eye blinking, etc.). It's better to leave hungry while the waters are still calm than to risk getting stuck in an emotional storm with all eyes watching.[2]

"You may want to bring something to offer a visual clue as to the time it will take to be served, like a digital clock. Breaking the wait up into more manageable chunks may also help. For example, telling your child you will be waiting ten minutes, then taking them outside for a quick walk, then coming back in and telling them you will be waiting [another] ten minutes for your food to come [can help]," says Tara Woodbury.

Lisa Bertuccio recommends bringing a go-to bag to both her neurotypical and neurodiverse clients. "This is just a bag with crayons, paper, fidget toys, playdoh and other small items that will keep the child entertained quietly or provide a way to help calm any anxieties."

DEALING WITH PICKY EATERS

Taylor Covington of *The Zebra* advises that

> many children with autism have a very limited diet because of the sensory issues surrounding eating. Brainstorm foods that are "safe" and

travel friendly. For instance, if your child enjoys macaroni and cheese, consider single-serve containers you can make in your hotel room's microwave or with water from a hotpot you pack. Keep a large stash of safe foods on hand, and research restaurants before you leave. And encourage food exploration. If your child knows you have safe foods, he might be willing to try just one bite of something new. The excitement of a new destination can make this more inviting.[3]

"Two weeks ahead of a trip to Disney, we fill out a form which gets submitted via email to their dietary department," recounts Jason Adlowitz. "Upon arrival at our Disney restaurant, we ask for a chef to come to the table and discuss with Bailey what items on the list he would like and what they have available to prepare for him. We've had amazing experiences with Disney. Some restaurants bend over backwards to provide the best service."

DON'T LET YOUR FOCUS STRAY

"Pay attention to what your [child] is doing at all times," warns Mark Hutten:

Younger children on the spectrum don't think twice about leaning over and stealing a few onion rings from the guy at the next table or staring-down the teenager in a nearby booth. Don't wait until Pepsi has been spilled all over your pants before asking that your youngster's drink be served in a "to go" cup with a lid. And if you attempt a dining out experience that fails miserably, just leave early and go home. Do not use the trip home as an opportunity to lecture your child. After he has calmed down, talk with your child about what worked, what didn't, and what everyone can do differently the next time you go out to eat. Rehearse this at home (i.e., play a game called "eating out"). Practice makes perfect—don't give up![4]

Not giving up has really been the lesson of all the previous chapters. Coming up, I'll discuss some travel modifications that aren't covered under the main categories of this book, as well as specific destination suggestions, including venues that feed into children's unique passions.

PARENT SPOTLIGHT—
DANA MARCINIAK, BUFFALO, NEW YORK

Dana Marciniak is the mother of one neurotypical child and a nonverbal son who has been diagnosed with autism. She usually travels with three children—they bring along their nephew so that their youngest child has someone to play with. Dana's main apprehension about traveling is "the waiting and anxiety involved—it keeps me up at night—making sure I have backup planes chosen for late or delayed flights. Also, people staring or looking at us because of Ian's stimming."

Yet she travels because she wants to give her children "exposure to [new] things and experiential learning. The main challenge is making sure each step of the travel experience is good for Ian! That can be exhausting when you're talking about a plane ride, luggage, Ubers, etc." She says the biggest misconception about traveling with children on the spectrum is "that we don't travel at all or that our kids don't enjoy traveling, and that only children [on the spectrum] cry or get anxious on flights."

Restaurant dining can be an issue, but Marciniak says that because she brings Ian to restaurants a lot, he understands the process. Her strategies show the importance of giving a child some decision-making power during the meal:

> I always bring an iPad, of course, and try to limit his time on the iPad before going out [so he's anxious to use it]. When we arrive, I ask for a booth or table along the perimeter of the restaurant, but I let Ian pick his own seat. When it's time to order, I pull up photos on my phone of the menu items I know he likes and let him choose. If I can create a picture menu beforehand, I will, but that's not always possible.

To reduce waiting, Marciniak says she orders an appetizer that she knows will come quickly or she asks for it to come out first:

> Sometimes, I will bring a small snack in my purse just in case (like fruit gummies or a Nutri-Grain bar). I'm prepared to walk around the restaurant while the food is being prepared or if we're waiting for the check. I have extra clothes in the car as his shirt sometimes doubles as a napkin. If

he needs to release some "energy" while we eat, I let him walk around our table a bit or sit on his knees, whatever is most comfortable for him.

Marciniak also tries to use the dining experience to expand her son's horizons but is aware of his limitations. "We let Ian try food from our plates to see if he'll try anything new. He does about 50% of the time. But when he's done eating, he's done. We have to move fast when he's ready to go, so . . . we bring cash so we can pay [quickly]."

Familiarity also helps smooth the dining process, she says:

Once I find a place that treats us/Ian well, we go back a lot, and the staff gets to know him. At a chicken wing restaurant here in Buffalo, they used to have his Sprite ready for him when he walked in after school with my mom and had his double order of medium chicken wings already in the fryer. The [people at the] pizza place around the corner watched our kids grow up and always ask about Ian. On his tenth birthday, the owner even invited him into the kitchen to make his own pizza!

Marciniak cites the greatest difference between her neurotypical and neurodiverse children on vacation is "explaining the travel process. My NT kiddo can be told it's gonna be a long wait and [he'll] complain, while my kiddo on the spectrum insists that the explanation does not make sense and the result is a highly stressful situation." The main tweaks she's made are "calling ahead to TSA and getting through lines faster. Calling ahead anywhere to reduce the wait time, really. Staying in homes or large hotel rooms so my kiddo on the spectrum feels like he can move about." One of her greatest difficulties: "Our kiddo is not potty trained so if it's just me with him, we have to find a family bathroom, or I have to bring him into the women's room . . . which could get us looks."

CHAPTER 15 ACTION STEPS

1. Preview the experience through role play before venturing out.
2. Do your due diligence: Know your child's triggers and avoid restaurants that could spark them. Check out patrons' experiences with a restaurant's autism friendliness at online review sites and on apps like AutismVillage.com.

3. Though they are still few and far between, seek out restaurants that cater to neurodiverse families. If not autism friendly, make sure they're at least kid friendly.
4. Check in advance if the restaurant offers highchairs with buckles to prevent escape.
5. Use the bathroom before leaving home or the hotel.
6. Start small and slowly, maybe just visiting a local restaurant for appetizers or dessert during their slowest periods.
7. Pick menu items in advance from online menus to ensure there will be food available your child will eat. If you have a fussy eater, ask if the restaurant will allow you to bring your own food. Knowing there's "safe" food available may encourage your child to taste a bite or two of something new.
8. Call in advance and explain your situation to the waitstaff. Have food pre-made and waiting to reduce the child's anxiety.
9. Work out seating based on the needs of your child. Buffets might be too overwhelming. Outdoor dining might work out better.
10. If you do go to a buffet, ask if items can be combined and prepared by the chef (bread plus cheese can equal a grilled cheese sandwich).
11. Offer visual cues like a digital clock to help the child break up waiting time into smaller chunks.
12. Bring items to distract and calm. When in doubt, an iPad and pair of headphones may help.
13. Make sure the restaurant takes your payment information in advance and let them know if there's a meltdown, you may leave abruptly.
14. No matter how engaging the adult conversation, pay attention to your child throughout the meal to avoid embarrassing situations.
15. If the outing fails, don't be afraid to try again.

16

OTHER SUGGESTED MODIFICATIONS

*"I think the most important thing to know about traveling with chil-
dren, either on the spectrum or not, is that we put too much pressure
on ourselves to have the perfect vacation.*

*"There will be hurdles no matter how much planning we do but
focusing on the good times instead of what went wrong will give you
the confidence to travel again."* – Tara Woodbury, Escape into
Travel, Leland, North Carolina

Many helpful hints for traveling with anxious, inflexible, or neurodiverse children don't necessarily fall under the category headings of
the previous chapters. That doesn't make them any less important. In this
chapter, you'll find modifications that didn't fit elsewhere in this book.

HIRING TEMPORARY SITTERS

If you don't have a nanny, consider hiring one that works with children
with special needs. Care.com and Destination Sitters (www.destinationsit
ters.com) are two of several companies that advertise such a service. You
might use a travel nanny who is based in your destination city or have one
accompany you through the entire journey—including the plane ride from
home—based on what they offer and what you can afford. If hiring someone located at your destination, ask if your child can meet the caregiver
via video conferencing in advance so the two can develop a good rapport
before the trip starts. And perform your due diligence—insist on speaking
with references before you hire.

When Yvonne Wonder opened Destination Sitters thirteen years ago, she was shocked at the lack of standards for temporary sitter organizations, so she decided to set the standards herself:

> As a mom, I thought it was crazy. We make sure every single one of our sitters is interviewed, CPR- and First-Aid-certified, background-checked, drug-checked (beforehand and randomly), had their personal and professional references checked, and we also look into their social media. If something doesn't add up, we dig further.
>
> Our sitters wear our customized polo shirt and ID badge and bring a bag of toys. To cater to those with special needs, we try to hire teachers, pre-med students, people who work in the disabilities industry. And we suggest clients plan way ahead. We ask probing questions like, "What are the child's triggers? How do you comfort them? What do you usually do to redirect them?"

Wonder says her company has offices throughout California, Las Vegas, and Orlando. Post-COVID, she hopes to reopen in New York, Nashville, and Seattle, with plans to eventually expand to thirty-five cities as well as international locales. But physical location of offices doesn't limit the company's scope. Destination Sitters offers travel sitters who fly to wherever they're needed and have accompanied families anywhere from the Hamptons to Japan. They also cater to international guests visiting the United States who tour for a few weeks around the country (they have sitters bilingual in English and at least one of six different languages), as well as sitters for wedding and meeting attendees. They've worked with the NFL, AMEX, various medical and charitable foundations, and more.

"Our sitters will move heaven and earth to ensure clients have a nice vacation and they deserve it, considering the stress they're under," she says.

Not every traveling parent uses a third party like Destination Sitters to watch their children, however. Some use sitters provided by their hotel or resort or bring family or other trusted individuals on vacation with them.

"Once we hired a nanny at Beaches-Ocho Rios, because they were autism certified," says Kristen Chambliss of League City, Texas. "It was only for two hours, and we completed a long questionnaire and made sure our other child was there. The nanny did a good job, but it made me nervous since I didn't know them. We didn't do it again and usually invite my parents to come help or just bring our kids everywhere with us."

Jennifer Hardy of Cruise Planners in Kent, Washington, advises clients to "bring along someone on vacation that their child already knows" for that very reason. "I have had clients bring along their ABA therapists/technicians or [trusted] family members. Most [special needs clients] I work with seem to prefer to have this type of longer-term, pre-established relationship with their sitters rather than a temporary hire," she says.

INSURING YOUR HOLIDAY

With children in general, and those on the spectrum in particular, you can never be sure of what the future may hold. Special needs travel insurance may help you with last-minute cancellations, thanks to more lenient and flexible coverage that suits the needs of neurodiverse families.

"For this type of protection, I would recommend a Cancel Anytime or Cancel for Any Reason policy," says Jennifer Hardy:

> While there won't be a policy specifically designed for families with a member on the spectrum, these are going to offer you the most flexibility when it comes to cancellation. At Cruise Planners, we offer a custom Cancel Anytime policy through Allianz which provides up to 80% back for reasons not covered by a standard policy. There are absolutely some major exclusions to be aware of, so it would be of great benefit for a family to discuss their coverage needs with their travel advisor in order to locate the right policy.
>
> When it comes to trip interruption, this is harder. It really needs to be for a covered reason. If a client was specifically interested in trip interruption, I would need to conference in a licensed agent with the insurance company to address that question.

Lisa Bertuccio of Vacation Your World in Otisville, New York, also offers Cancel for Any Reason insurance to travel clients but explains that the caveat to this type of coverage is that it must be added to your trip typically anywhere from twenty-four hours to two weeks after initial deposit, and once your trip has started, it no longer applies. "Because of this, there is trip interruption insurance that I recommend families on the spectrum consider. I would also urge families to investigate what type of coverage their credit cards offer, since many include trip interruption and cancellation as part of their policy. Another tip I suggest when shopping for travel insurance is asking if preexisting conditions apply."

Bertuccio adds, "Legally, travel agents are not allowed to interpret or evaluate benefits or conditions of travel insurance but we do have to offer it and give our clients the information for each policy so they can decide for themselves which policy would best suit their needs."

"When leaving the country, we take out extra travel insurance," says Michael Sokol, a parent from Longmeadow, Massachusetts. "We use NOMADS. Their rate fluctuates, so if your cruise is six months away, buy your insurance early. If you cruise in the spring, the weather is great, but come winter, it may not be so nice, and the travel insurance jumps up to a higher rate."

Regarding attractions, according to International Board of Credentialing and Continuing Education Standards (IBCCES) President Meredith Tekin, Certified Autism Centers (CACs) will typically have a reentry policy, so if families buy tickets to an attraction but end up having to leave early due to a meltdown or overload, they can use the ticket another day. These CACs are noted in the Accessibility Guide that comes with the IBCCES Accessibility Card (discussed in chapter 14). Be sure to ask about such policies when booking any CAC attractions of interest.

TRAVELING WITH SUPPORT ANIMALS

A percentage of families on the spectrum have welcomed dogs into their lives to provide parents with additional assistance inside and outside the home. Some of these canines are emotional service animals (ESAs), whose primary purpose is to provide unconditional love and a calming influence. Others are autism service dogs, trained to "gently interrupt self-harming behaviors or help de-escalate an emotional meltdown. For instance, it might respond to signs of anxiety or agitation with a calming action such as leaning against the child (or adult) or gently laying across his or her lap."[1]

Recently, a hard line drawn between the two types made flying with a service dog a little more difficult. By March 2021, most airlines began to enforce rules by the Department of Transportation that banned ESAs aboard the passenger cabins on flights, a practice that over a million travelers engaged in during 2018, oftentimes with poorly trained and misbehaving pigs, monkeys, and other livestock sitting beside them free of charge. Airlines adopting the embargo included American, Delta, JetBlue, and United.

Airlines that no longer recognize ESAs will treat companion animals as normal pets, which must comply with size, weight, and breed restrictions, and may be forced to travel in the cargo section, which can cost travelers as much as $175 each way.[2]

Amtrak also has restrictions concerning service animals, which you can read at https://www.amtrak.com/planning-booking/accessible-travel-services/service-animals.html. And cruise lines have restrictions too; you can find a quick roundup at https://allthingscruise.com/cruise-line-pet-policies,[3] but also double-check with your Certified Autism Travel Professional or the individual line before booking your sailing.

While other types of animals won't qualify, if your ESA is a dog, one workaround might be to have the dog designated as a psychiatric service dog (PSD). To qualify, the handler must have a mental impairment that substantially limits one or more major life activities such as (but not limited to):

- Anxiety
- Autism
- Learning disabilities
- Panic disorders
- Phobias
- Post-traumatic stress disorder
- Severe depression

However, the dog can only qualify if it's well-behaved in public and has been individually trained to perform tasks related to its handler's disability. It can't just be their presence as a stress-reducer that qualifies them as PSDs.

According to the ESA Doctors' website,[4] these tasks can include:

- Calming a handler having an anxiety or panic attack with pressure or tactile stimulation
- Helping ground and reorient a person having a psychotic episode
- Reminding a person to take their medication
- Preventing someone from oversleeping
- Interrupting obsessive-compulsive or self-destructive behaviors

All airlines flying to and from the United States are required to allow PSDs aboard their flights, and to allow them to fly in the cabin with their handler free of charge, as long as the accompanying passenger submits the Department of Transportation's Service Animal Transportation Form that attests to the dog's health and behavior, and certifies that their dog has been trained to perform tasks related to their handler's disability.

Due to privacy laws, airlines are not allowed to ask about the passenger's specific disability, and they cannot require passengers to demonstrate the task their PSD has been trained to perform. The form also requires the name of the PSD's trainer, which can be the owner (third-party training is not a requirement for PSDs). If the passenger makes knowingly false statements, they will be subject to fines and other penalties.

For more information, check the Department of Transportation's website at https://www.transportation.gov/individuals/aviation-consumer-protection/service-animals. For more information on assistance dogs in general and how and where to procure one, visit https://www.autismspeaks.org/assistance-dog-information.

TRAVELING WITH THE OLDER NEURODIVERSE CHILD

The good news is that traveling with neurodiverse children gets easier as they grow older, as long as you adjust your expectations and tweak your modifications, say travel agents who handle both neurotypical and neurodiverse clients.

"Usually by this age (over eighteen), they have situations managed better or have learned coping mechanisms to deal with [travel]," says Cathy Winter of EnVision Travel in Denver, North Carolina.

Their suggestions follow.

Plan Different

For success, travel advisors say that parents need to involve their older children both in the planning of trips as well as standing up for their rights when necessary. Many of these rights are explained on the Department of Transportation's website at https://www.transportation.gov/individuals/aviation-consumer-protection/traveling-disability.

"Make [older children] feel you value their opinion and thoughts. Ask them [what] places they would like to visit and plan around that destination. Getting them involved makes them feel important and a valued part of the family," says Nicole Graham Wallace of Trendy Travels by Nicole—Dream Vacations in McDonough, Georgia.

Tweak your own interactions with the older child, suggests Brenda Revere of C2C Escapes Travel in Bedford, Indiana. "Instead of having a special toy, depending on the type of autism, [provide] a journal for writing about the experiences, or a book about the destination."

"You have likely practiced having your child with high-functioning autism advocate for themselves as they have gotten older. Traveling is a good time for them to put those skills to use," says Tara Woodbury of Escape into Travel in Leland, North Carolina. "Make sure they are aware of their rights and how to calmly explain their needs to someone, especially if they're separated from you. At some point, depending on the severity of their autism, your adult child will want to travel on their own. You want them to have the tools they need," she says.

Parent Advocacy

Be aware that you may not have the same ability to guide and guard [your kids] as adults as you did when they were children, Woodbury warns:

> Because the ADA protects adults and children with ASD alike, the same accommodations that a child receives should [also cover] adults. However, legally you may have a fight on your hands. For example, at the airport, TSA generally wants everyone over the age of twelve to take off their shoes. This is when having a [note about special needs in the] PSS (Passenger Service System) helps.

"Once a child reaches the age of eighteen, you must be more careful as to rules related to flying, trains, boats, and other aspects of travel," adds Carl W. Johnson of Adventures in Golf (incorporated as Hamill Travel Services, Inc.) in Amherst, New Hampshire. "You may be required to show legal proof of guardianship over the person; check with your state agencies and with the applicable travel authorities prior to departure to ensure you have all that you need to legally address any possible situations that might arise."

Air Travel

With older children, "I still highly recommend TSA PreCheck," says Jennifer Hardy. "Not only do many airports have separate screening areas for these guests, making the wait times shorter, but the number of people in the waiting area is fewer. The TSA PreCheck screening experience is very consistent. This still can be combined with the TSA Cares program." (Read more about both of these programs in chapter 8.)

Hotels and Resorts

Unfortunately, many autism programs are geared toward young children, says Chandra Nims Brown of Cruise Planners—Global Travel Pros in Williamsburg, Virginia. "Finding locations and destinations that offer activities for older persons is important."

Jonathan Haraty of Adventure Luxury Travel in Whitman, Massachusetts, says:

> If the family member on the spectrum is over eighteen, options for parents become limited if they are looking for free time alone and still want [their] young adult in safe hands with well-trained and certified autism staff. Check [in advance to determine] if your destination is prepared to receive family members over eighteen on the spectrum. Beaches Resorts is one resort that provides one-on-one staff to entertain your family member while you are off [enjoying] activities or dining with other adult family members. (Read more about Beaches in chapter 12.)

At hotels and resorts, make sure you "have a contact so you do not have to continue to repeat your needs. Be sure that any [neurodiverse children over eighteen] are treated as adults even if their interests are different than neurotypical adults of their age," adds Markeisha Hall of Funtherapy Travel in Rosena Ranch, California.

Cruising

When it comes to cruise travel, anyone eighteen and over will not be able to participate in the onboard youth activities which may be designed for, or accommodating of, those on the spectrum. However, cruise line personnel may be able to make some accommodations for participation in other onboard activities, so be sure to ask, advises Hardy.

While on the Move

A primary alteration for older travelers on the spectrum would be to increase restaurant pre-planning and menu review, says Hardy. "Older children may still rely heavily on meals considered to be on the 'Kids' Menu,' so check the restaurants' self-imposed age restrictions and inquire about exceptions. It might help to seek out specific locations that have larger, adult versions of popular kids' meal items, such as pub-style food."

Having a medical power of attorney so you can advocate for your adult child is important. So is knowing in advance how to handle toileting issues, says Sarah Salter of Travelmation dba Sarah Travels Happy in Fayetteville, Georgia. "The [general] public may be less accommodating to special needs adults rather than 'cute kids.' Some destinations are not flexible with special needs regulations regarding access to supervision—like allowing parents to accompany their adult or older child [into the bathroom]—or with issues such as incontinence."

"For adults with incontinence, the installation of appropriate spaces to change and tube feed should be the norm and not just the exception," says Salter. "Just as family restrooms, wheelchair lifts in pools, and nursing rooms are becoming more common, so should spacious, climate-controlled areas with adult-sized changing beds."

Many parents of children with autism spectrum disorder (ASD) offer advice regarding toileting issues, including accompanying children of another sex into the toilet, at https://ourcrazyadventuresinautismland.com/using-public-restrooms-with-children-with-autism.

When in Doubt, Recruit Reinforcements

"Depending on their needs, it [might] be best to hire a caregiver that can alter travel plans [for the older child or adult with ASD] as needed without affecting the rest of the family," says Sarah Marshall of TravelAble in Midlothian, Virginia.

For High-Functioning Older Travelers Seeking a Little Independence

Frontier Travel Camp, Inc., is a summer camp alternative for high-functioning individuals aged sixteen and over. Offering twenty-four/seven supervision, Frontier and its staff, whom they describe as highly trained counselors, have spent the past twenty-plus years offering tours for individuals with varying learning disabilities, developmental disabilities, and/or other difficulties requiring specialized supervision and guidance. Owner Scott Fineman, who received his master's degree in clinical social work from Fordham University, advocates his tours as a way for neurodiverse travelers to experience independence, improve social skills, and increase self-esteem in a "secure and exciting" environment.

"Our typical travelers are those teens and adults who have been through the summer camp world and have exhausted their gains through it. They want to get out, make friends, see the world, and live life. What's important to them? Being away from their mom and dad, feeling like normal adults, being treated like average persons, and helping to choose what we do for the day," he says.

Fineman estimates that in the past few years, the proportion of neurodiverse travelers in his groups has grown to as high as 75 percent and his company enjoys an 85 to 90 percent repeat factor, which makes it easier for them to build trips around the needs of specific travelers. This is important, he explains, because of the difficult logistics in organizing such tours. "You're dealing with a million personalities. Take amusement parks. Some travelers can't stand loud noises, some love the excitement. It's the same with national parks—some love being around nature, but others are afraid. Fourth of July fireworks are something we've learned to avoid. We're constantly caring for each individual's different needs, which is why it's vital to have wonderful people working for you."

Frontier's groups can be as large as thirty persons, with one staff member for every three to four travelers. Since 1998, they have traveled extensively throughout the United States and Canada, including Hawaii and Alaska, as well as Europe with trips to Italy, England, Scotland, Ireland, France, Switzerland, Greece, Turkey, Malta, Australia, Scandinavia, Russia, and the Baltics.

Fineman groups his travelers by functioning level as opposed to age:

Our clients have higher levels of intelligence, good hygiene, and are better able to adapt and learn to be with other people. Most are verbal.

While they still need socialization norms and cues, we look to create a cohesive group of nice, kindhearted people. We don't accept those who have been hospitalized for physical violence, suicidal or sexual issues, severe trauma, or high levels of depression. And because of this, we haven't seen any behavioral issues on our tours.

For more information, contact Frontier Travel Camp at 866-750-2267 or http://www.frontiertravelcamp.com.

Some other companies offering travel for older children on the spectrum aged eighteen and older include:

- Exceptional Vacations, Inc.: 866-748-8747, https://www.exceptionaltrips.com
- Hammer Travel: 877-345-8599, https://hammertravel.org
- New Directions Travel: 888-967-2841, www.newdirectionstravel.org
- Sundial Special Vacations: 800-547-9198, http://www.sundialtour.com
- The Guided Tour: 215-782-1370, https://guidedtour.com
- Trips, Inc.: 800-686-1013, https://www.tripsinc.com
- Trips R Us: 508-405-0999, http://www.tripsrus.org

TIPS FOR THOSE WITH MOOD AND NEUROLOGICAL DISORDERS OFF THE SPECTRUM

While attention-deficit/hyperactivity disorder was covered in chapter 3, Tourette syndrome, bipolar syndrome, and borderline personality disorder (BPD) are three conditions that can also co-occur with ASD and require similar travel modifications. However, because they differ slightly, I've included some additional tips.

Tourette Syndrome

Tourette syndrome is an inherited neurological disorder with onset in childhood, characterized by the presence of multiple physical (motor) tics and at least one vocal tic; these tics characteristically wax and wane. It is a comorbid disorder with ASDs.[5] Tourette syndrome and other tic disorders affect more than 1 percent of school-age children across the United States, though that number is likely under-reported by as much as 50 percent because of lack of diagnosis. Tourette syndrome is three to four times more common in boys than girls.

Co-occurring conditions can be attention-deficit/hyperactivity disorder; obsessive-compulsive disorder; learning difficulties; behavioral issues such as aggression, rage, oppositional defiance, or socially inappropriate behaviors like shouting out slurs without provocation; anxiety; and mood problems such as depression or mania. Social skills deficits and sleeping problems might also co-occur.[6] In other words, these are many of the issues this book already tackles.

Much of the stigma that surrounds Tourette syndrome and tic disorders comes from misunderstanding. The Tourette Association of America produces informational cards with the heading "Why do I act this way? Because I can't control it" that help explain the effects of Tourette syndrome and tic disorders and are particularly useful in high-stress situations, like when traveling or in crowded places.[7]

Explaining the situation in advance to flight attendants and passengers sitting nearby might head off some of the confusion before it starts.[8]

Bipolar Disorder

Because bipolar disorder is a mood disorder rather than part of the autism spectrum, avoiding triggers is key. And of course, jet lag and changes of routine are two triggers that can occur during travel.

Along with the other tips offered for neurodiverse travelers, bipolar experts like Julie A. Fast advise prioritizing children's sleep (perhaps getting kids on a different pattern before you leave for a destination in a different time zone); staying at a hotel even when visiting friends so you can have better control over your surroundings; booking flights that mesh better with your schedule instead of opting for the cheapest available; bringing extra medicine in case of delays; and maintaining hygiene. They also suggest planning ahead for what might go wrong based on incidents in your past and how you successfully overcame them; getting adequate exercise, which always tends to elevate moods; realizing that mood swings may occur upon your return home and planning accordingly by scheduling a doctor's visit in advance; focusing on your children's needs, even if it means bowing out of a planned excursion or triggering conversation; maintaining flexibility; and, in times of panic, focusing on deep breathing.[9]

Borderline Personality Disorder

According to RTOR.org, BPD is a serious mental health condition characterized by difficulties in managing emotions effectively. The main feature of BPD is a strong pattern of instability in a person's relationships, self-image, or emotions. People who live with BPD can experience rapid changes in mood that last from a few hours to a few days. They may also experience issues of identity, self-harm (such as cutting), impulsivity, fear of abandonment, and feelings of emptiness.[10]

Kate, a blogger on the Deafinitelywanderlust.com website, says that she coped with her BPD on the road by using a journal to let out extreme emotions, and relying on self-soothing techniques she learned in dialectical behavior therapy, which taught her how to identify her emotions, evaluate whether they were appropriate for the situation, and act in ways that were effective and non-destructive. She focuses on her breathing to calm her racing thoughts.[11] Parents who wish to travel with their BPD children may wish to seek out dialectical behavior therapy training first. You can learn more about this therapy at https://childmind.org/article/dbt-dialectical-behavior-therapy.

MISCELLANEOUS STRATEGIES

Seeking Help When Planning

Know why and how to best look for and work with a travel advisor. This would be applicable to all families traveling, neurotypical or not. [Check with] Autism Travel (the IBCCES visitor site) to locate CATPs; the American Society of Travel Agents (ASTA) to find VTAs or Verified Travel Advisors; and Cruise Lines International Association (CLIA) to find those with ACC, MCC or ECC certifications. Personally, I have certifications from all three organizations, and they all offer something very different. CLIA is heavily cruise focused; a certificate in Accessible Travel is available but not yet searchable. ASTA is for general travel and code of ethics. These advisors truly have a commitment to the industry and have in-depth knowledge of all the associated laws, and this certification program is very intense. I would consider this to be the top certification for a travel advisor! And of course, the CATP from IBCCES has the focus on autism specifically. – Jennifer Hardy, Cruise Planners, Kent, Washington

It is always important to review different options and not to rush. You want to be able to pick the appropriate destination for the whole family to enjoy, so you may need to meet [with a travel advisor] several times before booking. – Suzanne Polak, AAA, Southington, Connecticut

Staying Flexible

Don't expect perfection. Acceptance of autism and other spectrum disorders does not mean giving up. There are so many benefits of traveling as a family such as sensory desensitization, educational benefits, life skills and of course, spreading autism awareness. Plus, there is always the benefit of enjoying time as a family. With the right preparation and planning, there is no reason families on the spectrum can't have an amazing getaway to the world's most beautiful and fascinating destinations. – Angela Romans, Rejuvia Travel, South Bend, Indiana

I know kids'/teen clubs are a big selling point at resorts and on cruise ships, and parents are thinking "Great!" I have asked mine (neurotypical and spectrum) to go the first day. If they don't like it, I don't force them. I think if you have a child on the spectrum, trying something new on vacation is not going to work well. – Starr Wlodarski, Cruise Planners—Unique Family Adventures, Perrysburg, Ohio

Choosing the Destination

If your vacation includes a destination with weather that is unlike what you experience at home, it is important to practice wearing the clothing and accessory items that will be needed [while there]. For example, if you intend to visit Alaska and your child is used to wearing shorts and tank tops, they may struggle with the transition to heavier, bulkier items and layers. Practice will help them adjust easier on the trip, and trialing will help the parents to know what new triggers may arise for children who may not normally experience clothing-related sensory issues. – Jennifer Hardy, Cruise Planners, Kent, Washington

Stay on Budget

It really helps to sign up for [programs offered through work] like Working Advantage or Plum Benefits for discounts. Loyalty clubs also offer points toward rentals or free dinners. Some credit cards give you miles to travel with. These all come in handy. Also, [don't ignore] the importance of getting travel insurance for trips planned way in advance. Most moms who have special needs children know how chaotic therapy/doctor's appointments scheduling can be. You may need to change or rearrange [vacation dates] and you don't want to lose all your hard-earned money. – Delight Dolcine, parent, Brooklyn, New York

We give our kids ways to earn and save money before the trip. It's a lot easier to say, "Do you have enough money for this?" instead of, "No, I won't buy it." – Kelli Engstrom, parent, Kent, Washington

Structure the Trip with the Child First and Foremost

A child who loves animals and enjoys interacting, learning, and watching them will benefit from a trip to the zoo. One who loves rockets and space exploration would enjoy a trip to Cape Canaveral. Switch the two and you'll have a bored, frustrated, and unhappy kiddo. – Chandra Nims Brown, Cruise Planners—Global Travel Pros, Williamsburg, Virginia

Communicate Your Needs and Stand Your Ground

If you ask for and receive an accommodation, and you need additional accommodations, they will not necessarily be offered; you need to request them. I can't stress this enough. [Suppliers] will absolutely work with you, but you need to ask for what you need, it will not be offered. You need to be an advocate for your family. – Michelle St. Pierre, Modern Travel Professionals, Yardley, Pennsylvania

Problems with travel suppliers? If you or your child is treated unfairly or discriminated against while traveling, do not pick a fight with the person you are in contact with. Instead, go to management to get help, as this is often more effective. Also, be [forearmed] with the company's disability regulations and accessibility rules, if possible.[12] – Taylor Covington, *The Zebra*

Toilet Issues

Prepare for toilet troubles. Children with autism, even if they are fully toilet trained, may struggle when out of their normal routine. They may find the public restrooms or hotel bathroom frightening and uncomfortable . . . because of all the sensory input of public restrooms. The loud hand dryers, automatic flushing toilets, unusual smells, and unusual sights can all add up to problems. Be prepared for these issues with noise-canceling headphones and sticky notes to put over automatic toilet sensors. Also, consider looking for family restrooms, which can be less frightening for your child. Practice visiting public restrooms before your trip to make this transition easier. And be prepared with extra pull-ups or changes of clothing in case a toileting accident happens.[13]
– Taylor Covington, *The Zebra*

If your child is not toilet trained, remember to pack your own sheets, plastic mattress covers, and choose hotels with self-serve laundry facilities.
– Anonymous mom

Activities

If your child is involved in an equine therapy program at home and you'd like to maintain that routine while traveling, check out the member stables at the Path International Member Center website (www.pathintl.org) or on the Autism Speaks website under Autism Friendly Services, Equine Programs (https://www.autismspeaks.org /resource-guide). Some of these member stables might accommodate a "drop-in" visit; call ahead to check. If your child just loves horses and doesn't need a therapy session, it's easy to google a horseback riding stable in the city or town you'll be visiting.

Aqua therapy might be easier to take on the road, especially since most destinations will have a pool. – Jennifer Hardy

And in General

When traveling over holidays like Christmas, expect long delays and flight cancellations due to weather. It's best to find a hotel and stay overnight prior to the day of travel and make sure you choose nonstop/ direct flights whenever possible. Always bring food and snacks. We have been on several flights that ran out of food due to long delays, etc., so always have food on hand. Finally, make sure that [kids'] clothing is layered and soft to avoid overheating, and discomfort. – Dawn Baeszler, AAA Travel, Lawrenceville, New Jersey

Prepare for the unexpected. If your child takes medicine on a regular basis, bring an extra weeks' worth and keep it on your person or in your carry-on while traveling. – Lisa Bertuccio, Vacation Your World, Otisville, New York

We always have a good day/bad day schedule for each day. And a plan of how to separate the kids if needed. – Kelli Engstrom, parent, Kent, Washington

Teaching your kids to be friendly to all children, no matter if they are in a wheelchair, have autism, or are just shy, can make a big difference in social situations.[14] – Nicole Thibault, Magic Storybook Travels, Fairport, New York

Because the Good News Is . . .

Vacations are possible for every family. If you dream of more than just the average vacation to the beach or mountains, we can make it work. Every family should be able to check vacations off their bucket list! – Olivia McKerley, Elite Events & Tickets, Grovetown, Georgia

ADVOCATE SPOTLIGHT: TRAVELABILITY AND THE STATE OF SPECTRUM TRAVEL

TravelAbility is an organization that wants *everyone* to enjoy travel. That's why their leaders have dedicated themselves to fulfilling a long-neglected need: teaching the travel industry how to adapt itself to better service those with physical and invisible disabilities. Regarding their mission as it affects the ASD community, Jake Steinman, founder and chief executive officer, explains: "The Americans with Disabilities Act (ADA) provides some baseline physical infrastructure for people with disabilities, and the Airline Carrier Access Act (ACAA) provides directions for air travel, but there is no ADA for information infrastructure and we are trying to fill that gap by educating hotels, attractions, and destinations about the needs of families with ASD or mood disorders when they travel and the barriers they face."

Their approach is multimodal and includes, but is not limited to, TravelAbility industry conferences, the *TravelAbility Insider* newsletter, a list of specialized travel providers, and their podcast series, *Accessible Destinations*. In addition, they've designed their website as a one-stop hub where destinations, hotels, and attractions can find the resources they need to become more accessible. To keep them on course, they've recruited an advisory board comprised of leaders in the disability travel community, media, and travel industry.

Steinman criticizes the current state of the travel industry for its

> lack of [supplier] training and awareness about how to prevent and react to meltdowns, tantrums, and other disruptive behaviors. . . . [Most] families with autistic children don't travel. Those that do find an autism-friendly destination, attraction, hotel, or restaurant and go back every year as they have the comfort of knowing what they will find. Personally, I believe the missed opportunity is for trained *cognitive experts* at major family destinations to offer "respite care" for autistic children so their parents can have a leisurely dinner out.

The future of spectrum travel will be propelled by those increasing awareness and providing greater access, says Steinman. He points to Mike DiMauro, who runs the Mike's Mission Facebook page (https://www.facebook.com/groups/mikedsmission), a "thirty-something with autism whose goal it is to visit all fifty states by car as he delivers Uber Eats in each to subsidize his trip." He also gives a thumbs-up to cities like Mesa, Arizona (www.visitmesa.com), and Visalia, California (www.visitvisalia.com), for becoming CACs, which means that 60 percent of their industry and city stakeholders have received training and passed certification exams given by IBCCES, the leading training organization. "Of course, in both cases the CEOs of those tourist boards have a child on the spectrum," he says.

Some of the suppliers he commends (aside from many discussed earlier in this book) include "Alaska Airlines' 'Fly for All' app, which has a video simulating the in-flight experience (learn more at https://blog.alaskaair.com/alaska-airlines/fly-for-all-app-ease-anxiety-of-air-travel); and Fun and Function (https://funandfunction.com), which provides equipment for sensory rooms in airports and hotels." He also recommends that travelers on the spectrum check out Threshold360 (https://map.threshold360.com), which provides software that one hundred thousand hotels, destinations, and conference centers have used to create videos to help future visitors learn and prepare for what they will experience. He also praises Infiniteach (https://www.infiniteach.com), which helps the travel industry develop "Know Before You Go" apps for travelers with ASD.

TravelAbility itself is taking a major role in enhancing accessible travel, both in its educational summits as well as at its pre-conference LaunchPad Pitchfest. This event throws a spotlight on both start-ups and established businesses developing assistive technology and products for the travel and disability community. Those with an interest in attending TravelAbility's summits and Pitchfest should check out their site at https://travelability.net/.

CHAPTER 16 ACTION STEPS

1. If you need support on the road, you can hire a temporary sitter at the destination or one that can travel with you from home. Be sure to interview in advance and thoroughly explain the special needs your child might have.
2. Look into travel insurance before you leave home. Cancel Anytime or Cancel for Any Reason policies exist, as do Trip Interruption policies, but you should be aware of exclusions. Travel agents can offer protection but cannot interpret or evaluate benefits.
3. If you travel with a support animal, airline rules changed in 2021. ESAs can no longer travel in most airline cabins; dogs must be designated as PSDs. Familiarize yourself with the requirements and restrictions.
4. Older children may still need support when traveling, but the rules are different. They may also prefer more independence. Several tour companies organize group tours for neurodiverse children over eighteen.
5. Conditions that run comorbid with ASD, such as Tourette syndrome, bipolar disorder, and BPD, have their own unique travel requirements.
6. Learn the designations for travel agents so you can pick the one that works best for you.
7. Stay flexible when traveling and don't expect perfection.
8. Don't assume that because a kids' club exists, it's the best choice for your child.

9. If the weather at your destination requires different attire than at home, practice wearing these garments before departure.
10. Take advantage of cost-savings whenever possible. Programs like Working Advantage and Plum Benefits can help. So can loyalty club points.
11. Don't assume special needs accommodations will be offered—you must ask for them. Advocate for your child's rights whenever necessary.
12. Understand how to make public toilets less frightening for children with ASD.
13. If your child is in equine therapy, look into local stables or programs at your destination.

Part V

DESTINATIONS AND SPECIAL INTEREST TRAVEL

17

SUGGESTED ITINERARIES

"Almost any destination can become possible with the right accommodations, planning, and the help of a knowledgeable travel planner." – Sarah Salter, Travelmation dba Sarah Travels Happy in Fayetteville, Georgia

For those with family members on the spectrum, the decision of where to travel is a crucial one. To play it safe, you may choose to go camping or patronize an all-inclusive resort or cruise ship that caters to the spectrum community, but if you decide to explore beyond those options, there are decisions to make and choices to consider. Fortunately, the world is becoming more autism friendly, and this could be your opportunity to take advantage.

Some locations, like the entire city of Mesa, Arizona, are becoming designated as Certified Autism Centers (CACs), so you can travel nearly anywhere and feel comfortable. Read more about Mesa later in this chapter, along with other cities that might not yet be certified but are increasing their outreach to the community. Please note that contact information for venues and attractions mentioned in this chapter can be found in other sections of this book, such as chapters 4, 12, 14, and 17.

Families may choose to build a vacation around their neurodiverse child's passion, be it trains and railways, dinosaurs, animals, woodworking, rocks—the list is as varied as the diversity of interests. Google is your friend. It's easy to delve in and discover which museums and venues cater to those interests and then build a vacation around them, using vendors and techniques from other areas of this book. I've included many examples in this chapter.

I asked parents and travel advisors for their suggestions of the best itineraries for those on the spectrum. Some, like Michelle St. Pierre of Modern

Travel Professionals of Yardley, Pennsylvania, were clear that it's a difficult question to answer without meeting the family first:

> I may be the odd man out on this, but the very nature of the spectrum means every child I've planned travel for is so different and I can't think of anything that's "cookie cutter." If you asked me about my favorite itineraries for kids (assuming you meant neurotypical) aged three to four, I would absolutely have a response, but for kids on the spectrum I need to talk to the parents, understand their concerns, understand any safety issues, etc.

St. Pierre, like most of the Certified Autism Travel Professionals surveyed, gave high marks to Beaches Resorts and Royal Caribbean Cruise Lines because of their commitment to autism-friendly programming. Others spoke well of theme parks like Disney. But parents shouldn't feel confined to only visiting those destinations that are autism friendly or autism certified. Here are the experiences (and lessons learned) by some parents who have ventured further afield:

> [On a trip to Paris,] we took the metro—bad idea—even though we were with a guide. It was loud and congested, and both kids had a meltdown. In general, cities and crowds were not easy and posed difficult challenges, so my tip for planning a city trip is to take lots of breaks, don't overschedule sightseeing, and make sure you work with your hotel concierge on choices of activities. My son was served fish with the head still on it, and he threw up at the table, so food choices are different than here in the States.
>
> On a trip to Alaska—by far the best trip the kids have taken to date—the dogsled was a big hit, both kids were receptive to the quiet and the calm, and fascinated by the glaciers and animals. They love to ski, and they love places that offer water. Nature in general appeals to them. Hiking also seems to work. Sailing on a large cruise ship was not great, the kids would have done better on a smaller ship.
>
> In Puerto Rico, swimming with dolphins was an amazing experience that worked so well. – Dawn Baeszler, AAA Travel, Lawrenceville, New Jersey

When you find something that works, stick with it. My children loved camping in Hungry Horse, MT, and hiking in Glacier National Park last summer, so we're basically repeating the same trip this year. We also visit the same places over and over because they're places that we all enjoy, and the familiarity means less anxiety for both my neurotypical kid and my ASD kid. – Adrienne Query-Fiss, parent, Seattle, Washington

The purpose of this chapter is to help parents create their own itineraries. In the remaining pages, I've covered destinations where the entire city has declared itself autism friendly or has been designated as a CAC. I've also described destinations that, while not autism friendly in their entirety, offer enough attractions for spectrum families to warrant a visit; suggested vacations built around reportedly therapeutic sports such as golfing, adaptive skiing, and scuba diving; and then finally, provided a list of museums and venues that cater specifically to special interests.

AUTISM-FRIENDLY DESTINATIONS

With the increased awareness of autism spectrum disorder (ASD) over the past several years, some entire cities have devoted funds and energy to become autism friendly. It makes sense: If a hotel is going to cater to those with special needs, then why not encourage the surrounding restaurants, venues, and recreational facilities to do the same? Inclusive destination marketing will hopefully be the wave of the future.

Mesa, Arizona, located next to Phoenix, was the first city to become a CAC, and Visit Mesa, the city's marketing arm, is quick to remind potential visitors with ASD of their nationally recognized status as "one of the most family friendly communities in the United States. Mesa hotels and attractions are rolling out the red carpet to welcome travelers on the spectrum. Whether creating a work of art, panning for gold, or exploring our stunning desert landscape for the first time, Mesa offers a full-sensory experience for every family."[1] A visit to their website at VisitMesa.com offers an extensive list of local autism-certified hotels, dining options, attractions, and activities to enjoy, along with certified sensory guides where available.

If you're visiting Mesa but also interested in watersports and other outdoor activities, Lake Havasu is only about three hours away. Go Lake Havasu, the marketing organization for Lake Havasu City, is also designated as a CAC. Their page on the Autism Travel website (https://autismtravel.com/directory/go-lake-havasu/) explains how neurodiverse families can best enjoy "Arizona's Playground."

Also in Arizona, the town of Queen Creek's Park and Recreation Department has recently been designated as a CAC, as have many venues in the city of Tempe.

Arizona is not the only autism-friendly game in town. Destinations featuring a preponderance of ASD-friendly resorts, restaurants, and activities are actively spreading the word. For example, Visit Visalia (in California) is now a CAC, and since 2018, Myrtle Beach, South Carolina, has been working to extend autism-awareness citywide. One outcome is that the Myrtle Beach Parks, Recreation & Sports Tourism Department was designated as a CAC. Another result is the Champion Autism Network (CAN) card, which allows families to alert travel vendors to their child's diagnosis. It's available free to those who request it at the Myrtle Beach Welcome Center.

What are some benefits of the CAN card? According to their website at championautismnetwork.com, when going to a participating restaurant, "show your CAN card to the host or server. Without a word, these restaurants will understand where to seat you and that 1) you may have to leave at a moment's notice; 2) you may be bringing in food for a member of your family; 3) you may have special requests when ordering; and 4) your family member with autism may have trouble sitting still, may not make eye contact, and may be nonverbal."[2] In addition, families can use the CAN card at crowded attractions to skip the line, get private seating or expedited service at a restaurant, and for curbside check-in at hotels.

In 2016, the Town Council of Surfside Beach, South Carolina, less than twenty miles from Myrtle Beach, voted to issue a resolution declaring the town the world's first Autism-Friendly Travel Destination.[3] Now certified by CAN, it continues to work toward being a judgement-free, ASD-friendly location, where businesses host events specifically geared to neurodiverse kids.

Many destinations in Colorado market autism-friendly activities. For example, resorts such as Crested Butte, Beaver Creek, Breckenridge, Keystone, and Vail offer adaptive ski programs and ASD-friendly lessons during the winter months. (A list of adaptive skiing venues follows later in the chapter; many operate ASD-friendly outdoors programs and camps during the summer months as well.)

In Florida, aside from Disney World, Orlando offers spectrum-friendly CAC attractions like Discovery Cove, Island H2O Live!, and SeaWorld, along with several CAC and autism-friendly hotels and vacation rentals. For those who want to vacation on both land and sea, Orlando is fairly close to the cruise ship departure ports of Port Canaveral (fifty-three miles) and Tampa (eighty-four miles), whereas Fort Lauderdale (217 miles) and Miami (235 miles) require a slightly longer journey.

Though not marketed as such, even big cities, despite their noise and crowds, can be acceptable spectrum-friendly destinations. Take Pennsylvania. In Philadelphia, you have CAC attractions such as the Museum of the American Revolution and the autism-friendly Touch Me Museum and Philadelphia Zoo as well as the quirky, if not necessarily autism-friendly, Mutter Museum, featuring body parts preserved in formaldehyde and its skull collection. (My kids still rave about it!) Then it's only thirteen miles to the LEGOLAND® Discovery Center Philadelphia in Plymouth Meeting (AF); twenty-two miles to the Elmwood Park Zoo—the first zoo in the world designated as a CAC; twenty-five miles to Valley Forge; twenty-seven miles to Sesame Place in Langhorne; sixty-four miles to Dorney Park in Allentown; and ninety-five miles to Hershey's Chocolate World Attraction. All are CACs or autism-friendly venues.

Further south in Texas, Houston features a Kid's Sensory Suite at the Omni Hotel and several CAC attractions such as Space Center Houston and the Houston Museum of Natural Science, while the Children's Museum (AF) offers sensory friendly days. About four hours away by car, the Fort Worth Zoo, with its seven thousand animals, is considered one of the finest zoos in the United States and is also a CAC.

With two large CACs, namely the Bloomington-based Mall of America and one of its attractions, Nickelodeon Universe, Minnesota hopes to attract its share of those traveling with ASD. Other Twin Cities local attractions catering to the population include the ValleyFair Amusement Park in Shakopee, Sensory Sundays at the Walker Art Center, and a sensory friendly visitor information center at the Como Zoo in St. Paul.[4]

Finally, don't forget about New York City. Yes, it's considered to be the center of the universe by some, attested to by its hordes of annual tourists. But consider this: you can stay in an Airbnb or vacation rental in the nearby suburbs—your child might relish the excitement of commuting into the city every day by train, bus, or ferry (it could become his or her holiday highlight!). In town, you can enjoy the convenience of hundreds of restaurants that will serve or deliver whatever cuisine you crave. Meanwhile, the city's bevy of autism-friendly museums and attractions cater to every unique circumscribed interest. A New York City–based holiday may seem counterintuitive to families seeking quiet, but it might be worth a shot.

GOLF VACATIONS

If Billy Mayfair's ASD diagnosis didn't stop him from defeating Tiger Woods in a PGA Tour playoff, maybe it's a sport to consider for your child. Even better, it's an activity that also can provide more travel possibilities than can be mentioned in these pages.

The golf veteran went public in 2021—after receiving a diagnosis resembling Asperger's syndrome in 2019 at the age of fifty-two—which informed him of why he'd long self-identified as "obstinate and defiant." In an interview, he explained to *Sports Illustrated* how the diagnosis clarified issues that had puzzled him in the past, such as why he'd had issues adapting to change, passing timed exams, and understanding how to respond to affection.[5]

Despite all that, Mayfair is a veteran on the PGA Tour. While your child might not achieve tournament status, learning golf could lead to many pleasant family travels together around the world.

A story on NJ.com featuring Jen Hong of the #EGameON autism golf program in Jupiter, Florida, explains why. Says Hong: "People with autism, in general, do very well with repetitive tasks, with doing things that are structured. They're able to work on their swings over and over again, and if you have your swing down, you'll have 'muscle memory' and be able to succeed more."[6]

In addition, while golfers play in foursomes or on teams, golf is ultimately an individual sport, the type usually recommended for those on the spectrum. According to Aaron Hensley, general manager of the Oak Valley Golf Course, the first golf course in California to become a Certified Autism Center: "Unlike team sports that require a player to react to or even anticipate a teammate's actions, golf is a game of one. That makes it ideal for autistic children who have difficulty reading the social cues of those around them."[7]

There are more and more golfers on the spectrum as well as many who have found playing golf to be both soothing and satisfying, says Carl W. Johnson of Adventures in Golf (incorporated as Hamill Travel Services, Inc.) in Amherst, New Hampshire. "Golf courses are quiet, [with] not many people near you, nothing truly unexpected will happen (no fire alarms or other unexpected noises). Truly, golf is one of the only sports which can apply to anyone of nearly any skill level. Almost any limitations can be overcome."

A twenty-eight-minute documentary, *Voices from the Outside*, available at Voices From The Outside on Vimeo (https://vimeo.com/212905669), highlights how golf can help kids with autism. The documentary profiles three families who are part of the Brooklyn Junior Autistic Golfers Academy, which provides interactive golf lessons for children on the autism spectrum. The Brooklyn Junior Autistic Golfers Academy (https://bjaga .net) was established in 2010 with a non-profit 501(c)(3) status and identifies as an organization dedicated to working with all families, especially those with youth who are living with autism.

If children express interest, it might be worth taking a lesson or two; if they enjoy it, it can lead to a lifetime of golf vacations together.

After studying recommended "family-friendly" courses, I discovered that all offer one or more of the following:

- The course is located at a family-friendly resort or one close to theme/water parks or other child-friendly activities.
- There's instruction offered that's specifically geared to kids.
- There are nine-hole courses, which makes for a shorter game where attention is less likely to wane. "Walking-only" courses remove the fear of a cart colliding with an eloping child.
- The course offers "kid tees" (also called family tees, junior tees, or short courses) where tees are moved closer to greens or in the middle of fairways, heightening the opportunity for success.
- There's special pricing in place that makes it affordable for children to give it a try. Often these discounts are offered in conjunction with an adult playing alongside and paying full price.
- The course offers clubs sized specifically for junior players, often provided free or at a discounted rate.
- There are opportunities for families to bond, such as family tee times, instruction, and tournaments.
- There's a variety of practice opportunities aside from the course, such as putting greens, chipping areas, and driving ranges.

If you're sold on the idea of family golf vacations, *Golf Digest* lists one hundred great golf courses for families at https://golf.com/travel/best-golf -resorts-families-top-100-resorts/.[8]

We asked Carl W. Johnson for his staff's top family-friendly locations with golf onsite or nearby. Here are their picks (in alphabetical order by state) and their rationale as to why these properties made the list:

- Colorado
  - ○ **The Broadmoor**
    1 Lake Ave, Colorado Springs, CO 80906
    844-602-3343
    https://www.broadmoor.com
    Historic resort with two golf courses onsite. Lots of outdoor activities locally including hiking, biking, Pikes Peak Railway. Denver is nearby as well, as is the Air Force Academy and Olympic Training Center.

  - ○ **Vail**
    https://discovervail.com/things-to-do/golf
    While known as a popular skiing village, the greater Vail/Beaver Creek area offers plenty of summer activities as well. Hiking, biking, golf, fishing, etc., are all available near the various skiing villages. Dining and shopping options are also abundant. Variety of accommodations options exist including multi-room villas, etc.

- Florida
  - ○ **Greater Orlando including the Walt Disney complex and Lake Buena Vista**
    https://www.visitorlando.com/search/?q=golf
    Known as one of the easiest areas to access from any part of the United States, with plenty of things to see and do within thirty to sixty minutes. Water parks, theme parks, shopping, fishing, and tons of golf locally. Walt Disney's facilities always cater to families so nearly everything will be family friendly.

- Georgia
  - ○ **Sea Island Resort**
    100 Cloister Dr, Sea Island, GA 31561
    877-661-0836
    https://www.seaisland.com

Wonderful beach location with three courses onsite—one is an annual PGA Tour stop. Known for their Beach Club which caters to children of all ages. Always known as a family-friendly resort. Mostly water activities—fishing, boating, swimming, etc.

- Missouri
  - **Big Cedar Lodge**
    190 Top of the Rock Road, Ridgedale, MO 65739
    800-225-6343
    https://bigcedar.com
    Plenty of family activities onsite: fishing, boating, hiking, biking, underwater-themed bowling alley, explorable caves, plenty of golf options onsite with a new Tiger Woods course, etc. Variety of accommodations options as well including multi-room cottages, etc. This is regarded as a truly family-focused destination.

- New Hampshire
  - **Omni Mount Washington Resort**
    310 Mount Washington Hotel Road, Bretton Woods, NH 03575
    603-278-1000
    https://www.omnihotels.com/hotels/bretton-woods-mount-washington
    Historic resort, known for its stunning views. Outdoor activities are what draws most people here: hiking, biking, fishing, and visiting Mount Washington by roadway, hiking, or railway. Locally, there is North Conway shopping. Skiing in the winter months. Championship golf onsite designed by Donald Ross.

- South Carolina
  - **Kiawah Island Golf Resort**
    1 Sanctuary Beach Drive, Kiawah Island, SC 29455
    800-654-2924
    https://kiawahresort.com
    Wonderful beach location with five courses onsite—the Ocean Course is world-ranked and a past "major" tournament venue. Small water park at main hotel. Overall, it's known as a kid-friendly resort. Charleston is one hour away for anyone wanting a historic city tour.

- Texas
  - **Horseshoe Bay Resort**
    200 Hi Circle North, Horseshoe Bay, TX 78657
    877-611-0112
    https://www.hsbresort.com
    Offering four golf courses onsite and a real grass miniature golf course, three pools, lake activities (kayak or stand-up paddleboarding), plus a Jungle Kids' Club with rock climbing walls, indoor slides, and children's art classes, this Texas property offers everything you might want in a family getaway, only an hour outside Austin.

ADAPTIVE SKIING

Individual sports like golf, cycling, track, and skiing have long been advocated for children on the spectrum because they can participate without stress-filled social interaction. To that end, many ski resorts have started using adaptive skiing techniques to help children with developmental challenges as well as physical ones. Adaptive skiing programs are tailored to the individual: while one skier may require special equipment to enjoy the sport, another might depend on behavioral supports. Instructors understand how to incorporate children's obsessions into the lesson itself and use cognitive reframing to get points across.[9]

In her blog for Ski Utah, Jodi Saeland lists several tips for parents who want to introduce their children to skiing, offered to her by Alan Lindsey, an instructor with the National Ability Center. These include the following:

1. When you call to schedule a lesson, make sure the ski school knows your child has a physical or learning disability. That way, they can make sure a specialized instructor works with your child.
2. Talk with the instructor when dropping the child off. Give them some tips on how best to communicate with your son or daughter.
3. If your child has a special interest, let the instructor know as it provides a talking point that may help the student and instructor bond.
4. Don't set expectations. Your child may stay on the beginner hill all day and that is okay, or they may move quickly to the higher lifts.

Sometimes the learning curve is slower and as a parent, you don't want to push your child.[10]

Here are some suggestions for ski resorts that offer adaptive skiing programs for both physically challenged and spectrum families; many of them offer summer programs as well. Consider all to be "AF" or Autism Friendly. Addresses may refer to company's main address and not the ski resort/slope, so double-check for an exact location before heading out.

- Alaska
  - **Challenge Alaska Adaptive Ski and Snowboard School**
    426 Crystal Mountain Road, P.O. Box 1166, Girdwood, AK 99587
    907-783-2925
    http://www.challengealaska.org

- California
  - **Bear Mountain: United States Adaptive Recreation Center**
    43101 Goldmine Drive, Big Bear Lake, CA 92315
    909-584-0269
    http://usarc.org

  - **Lake Tahoe: Achieve Tahoe**
    2680 Alpine Meadows Rd, B3686, Alpine Meadows, CA 96146
    530-581-4161
    http://achievetahoe.org

  - **Northstar at Tahoe**
    (Many lodges with different addresses)
    800-466–6784; Disabled Sports at 530-581-4161
    https://www.northstarcalifornia.com/

  - **Mammoth Mountain, California: Disabled Sports Eastern Sierra**
    Post Office Box 7275, Mammoth Lakes, California 93546-7275
    760-934-0791
    https://disabledsportseasternsierra.org

- Colorado

  Ascendigo: Skiing Lessons and Programs in Aspen Snowmass, Summer Camp in Roaring Fork Valley
  Main Office: 818 Industry Place, Suite A, Carbondale, CO 81623
  970-927-3143
  http://www.ascendigo.org/adventures-camps
  Note: Ascendigo Autism Services' Adventures has reportedly developed the first certified ski teaching program specifically for individuals on the autism spectrum.

  ○ **Beaver Creek Resort**
  26 Avondale Lane, Avon, CO 81620
  970-754-4636
  https://www.beavercreek.com/plan-your-trip/ski-and-ride-lessons/category/adaptive.aspx

  ○ **Challenge Aspen at Snowmass**
  P.O. Box 6639, Snowmass Village, CO 81615
  970-923-0578
  https://challengeaspen.org

  ○ **Crested Butte, Adaptive Sport Center**
  19 Emmons Road, Mt. Crested Butte, CO 81225
  970-349-2296
  https://www.adaptivesports.org/

  ○ **Ignite Adaptive Sports at Eldora Ski Resort**
  2861 Eldora Ski Road #140, Nederland, CO 80466
  720-310-0328
  https://igniteadaptivesports.org/

  ○ **Keystone Adaptive Center at Keystone Ski Resort**
  P.O. Box 697, Breckenridge, CO 80424
  970-453-6422
  https://boec.org/keystone-adaptive-center

  ○ **Steamboat Adaptive Recreational Sports (STARS)**
  35465 US Hwy 40, Steamboat Springs, CO 80487
  970-870-1950
  https://steamboatstars.com

- ○ **Telluride Adaptive Sports**
  568 Mountain Village Blvd, Suite 101 (P.O. Box 2254), Telluride, CO 81435
  970-728-5010 (general); 970-728-3865 (program)
  https://tellurideadaptivesports.org

- ○ **Vail Resorts**
  P.O. Box 7, Vail, CO 81658
  970-754-3264
  https://www.vail.com/plan-your-trip/ski-and-ride-lessons/category/adaptive.aspx

- ○ **Winter Park Resort: National Sport Center for the Disabled**
  33 Parsenn Road, Winter Park, Colorado 80482
  970-726-1518
  https://coloradoskiauthority.com/winter-park/disabled-skiing

- • Idaho
  - ○ **Higher Ground**
    160 7th Street W, P.O. Box 6791, Ketchum, ID 83340
    208-726-9298
    https://www.highergroundusa.org

- • Maine
  - ○ **Maine Adaptive at Sunday River Resort**
    P.O. Box 4500, 15 South Ridge Road, Newry, ME 04261
    800-543-2754
    https://www.sundayriver.com/maine-adaptive

  - ○ **Horizons Skiing—Carrabassett Valley and Locations around Maine**
    3000 Outdoor Center Rd, Carrabassett Valley, ME 04947
    Also: 675 Old Portland Rd, Brunswick, ME 04011
    207-237-2676
    https://www.adaptiveoutdooreducationcenter.org

- • Montana
  - ○ **DREAM Adaptive Recreation, Whitefish**
    P.O. Box 4084, Whitefish, MT 59937
    406-862-1817
    http://www.dreamadaptive.org

○ **Eagle Mount at Bridger Bowl**
15795 Bridger Canyon Road, Bozeman, MT 59715
406-586-1781 (Eagle Mount) or 406-587-2111 (Bridger Bowl)
https://bridgerbowl.com/lessons/adaptive-programs

• New Hampshire
○ **New England Disabled Sports**
39 Loon Brook Road, Lincoln, NH 03251
Loon Mountain: 603-745-9333
Bretton Woods: 603-278-3398
http://nedisabledsports.org

○ **Mt Attitash Mountain Resort**
Base area: 775 Route 302, Bartlett, NH 03812
800-223-7669
https://www.attitash.com/adaptive-programs

○ **Cannon Mountain Adaptive Sports Partners of the North Country**
297 Main Street, P.O. Box 304, Franconia, NH 03580
603-823-5232
https://www.cannonmt.com/ski-ride/adaptive-sports

○ **Gunstock Mountain Resort Lakes Region Disabled Sports**
719 Cherry Valley Rd, Gilford, NH 03249
603-293-4341 (Resort) or 603-737-4365 (Lakes Region Disabled Sports)
https://lradaptive.org

○ **Granite State Adaptive at King Pine at Purity Spring Resort**
Purity Spring Resort: 1251 Eaton Road, Route 153, Madison, NH 03849
Granite State Adaptive: 44 Mirror Lake Drive, Mirror Lake NH 03853
603-367-8896 (Resort) or 603-387-1167 (Granite State Adaptive)
https://www.gsadaptivesports.org/snowsports.html

- ○ **Mount Sunapee Resort's New England Healing Sports Association**
  P.O. Box 2135, 180 Mount Sunapee Rd, Newbury, NH 03255
  603-763-4400 (NEHSA) or 603-763-9158 (adaptive winter programs)
  https://nehsa.org/winter-programs

- ○ **Waterville Valley Adaptive Sports**
  1 Ski Area Road, Waterville Valley, NH 03215
  800-468-2553
  https://www.waterville.com/wv-adaptive

- New Mexico
  - ○ **Ski Apache Adaptive Sports**
    1286 Ski Run Road, Alto, NM 88312
    575-937-6954
    http://skiapacheadaptivesports.com

  - ○ **Ski Santa Fe and Albuquerque Adaptive Sports Program**
    P.O. Box 5676, Santa Fe, NM 87502
    505-570-5710
    https://www.adaptivesportsprogram.org

- New York
  - ○ **Adaptive Sports Foundation at Windham Mountain Resort**
    P.O. Box 266, 100 Silverman Way, Windham, NY 12496
    518-734-5070
    https://www.adaptivesportsfoundation.org

- Oregon
  - ○ **Oregon Adaptive Sports**
    63025 OB Riley Road, Suite 12, Bend, OR 97703
    541-306-4774
    https://oregonadaptivesports.org/sports

- Pennsylvania
  - ○ **Pennsylvania Ctr for Adaptive Sports at Camelback Mountain, Tannersville**
    4 Boathouse Row, Kelly Drive
    Philadelphia, PA 19130
    215-765-5118
    https://www.centeronline.com

- **Two Top Mountain Adaptive Sports Foundation**
  10914 Claylick Road, Mercersburg, PA 17236
  717-507-7668
  https://www.twotopadaptive.org

- **Liberty Mountain Resort**
  78 Country Club Trail, Carroll Valley, PA 17320
  717-642-8282
  https://www.libertymountainresort.com/plan-your-trip/ski
  -and-ride-lessons/category/adaptive-programs

• Utah
  - **Common Ground Outdoor Adventures at Beaver Mountain**
    335 N. 100 East, Logan, UT 84321
    Logan Office: 435-713-0288, Beaver Mt. Office: 435-946-3205
    http://www.cgadventures.org

  - **National Ability Center at Park City**
    1000 Ability Way, Park City, UT 84060
    435-649-3991
    http://www.discovernac.org

  - **Adaptive Sports at Snowbasin Resort**
    3925 E. Snowbasin Road, Huntsville, UT 84317
    888-437-5488 (Resort) or 801-695-7074 (Ogden Valley Adaptive Sports)
    https://www.snowbasin.com/lessons-rentals/adaptive-sports

  - **Wasatch Adaptive Sports at Snowbird Ski Resort**
    9385 S. Snowbird Center Drive, Snowbird, UT 84092
    801-834-0476
    www.wasatchadaptivesports.org
    Note: Lessons at other Utah resorts, including Alta, Sundance, Solitude, Brighton, and Deer Valley, can be requested and will be provided based on availability.

- Vermont
  - **Bart Adaptive Sports Center at Bromley Mountain**
    3984 Vermont Rt. 11, Peru, VT 05152
    802-824-5522
    https://www.bromley.com/lessons-and-rentals/snowsports-school/adaptive-lessons-the-bart-center

  - **Adaptive Sports at Mount Snow**
    39 Mount Snow Road, West Dover, VT 05356
    802-464-4069
    https://adaptiveatsnow.org

  - **Smugglers' Notch**
    4323 Vermont Route 108 South, Jeffersonville, VT 05464
    802-644-8502
    https://www.smuggs.com/pages/winter/kids/adaptive-programs.php

  - **Vermont Adaptive Ski and Sports at Pico, Bolton, and Burlington**
    Vermont Adaptive Ski and Sports at Pico Mountain, Bolton, and Burlington
    P.O. Box 139, Killington, VT 05751
    Pico: 802-786-4991 ext. 25; Bolton & Burlington, ext. 27
    https://www.vermontadaptive.org

  - **Vermont Adaptive Ski and Sports at Sugarbush**
    P.O. Box 139, Waitsfield, VT 05673
    802-786-4991 ext. 26
    https://www.vermontadaptive.org

- Virginia
  - **Wintergreen Adaptive Sports**
    P.O. Box 4334, Charlottesville, VA 22905
    434-277-3408
    https://www.wintergreenadaptivesports.org

- Washington
  - **Outdoors for All**
    6344 NE 74th Street, Suite 102, Seattle, WA 98115
    206-838-6030
    https://outdoorsforall.org

- West Virginia
  - **Challenged Athletes of West Virginia (Snowshoe)**
    1 Snowshoe Dr, Snowshoe, WV 26209
    304-572-6708
    https://www.adaptivesports.org/adventures/winter-adventures
    /snowshoe (link for the Adaptive Sports Center, not the organization CAWV)

- Wyoming
  - **Adaptive Ski Program at Jackson Hole Mountain Resort**
    3395 Cody Ln, Teton Village, WY 83025
    307-733-2292
    http://www.jacksonhole.com/adaptive.html

Many thanks to SpecialNeedsTravelMom.com for compiling the foundation of this list, which I verified and built upon with further research. To read mini reviews of many of these resorts as well as links to some Canadian adaptive ski programs, check out their blog at http://special needstravelmom.com/blog/ski-resorts-adaptive-skiing-programs.[11]

DIVING IN: SCUBA VACATIONS

Swimming has long been considered a calming sport for those on the spectrum; it's something that can be done independently and

> being in the water often gives a sense of calm to individuals on the spectrum, especially children. The buoyancy and pressure of the water creates a supportive environment where they are able to develop their sensory processing skills. . . . On top of this, when children are able to play and have fun in the water, the positive effects are seen long after. Children with autism who spend time in the water are reported to be more talkative, social, cooperative, in a better mood, and able to focus better afterward.[12]

But what about diving? Non-profit organization Diveheart is one of a growing number of organizations dedicated to helping kids and adults learn scuba to build confidence and independence. According to Jim Elliott, the owner and founder, the training differs from that given to neurotypical students because it involves a larger, more specially trained team.

"The typical adaptive dive team consists of three to four divers," he says. "That includes the adaptive diver. It may also include a special person accompanying the adaptive dive team who is familiar with the adaptive diver and helps [him or her] stay focused through special communications like eye blinks or special hand signals. Someone with autism typically has two buddies on their team, but some with special conditions may require [more]."

Diving students on the spectrum are thriving. Take Nate Blumhoefer, a Minnesota-based diver, who says, "I wanted to be a diver for as long as I can remember. I view divers in the same way I view astronauts and cosmonauts. We explore places others only dream of." He reports that wearing his wet suit is one of his favorite aspects of the sport. "Being autistic, I need deep pressure to feel comfortable. I wear compression garments at all times, and wetsuits are the ultimate compression garment!"[13]

Diveheart's Elliott explains why those like Blumhoefer are so enthusiastic about the sport. "Pressure increases underwater, and that ambient pressure is soothing and comforting for those with disabilities. Underwater also allows those with autism to escape surface distractions and triggers. Plus, the cool factor is off the charts and helps those with autism self-identify as someone who is a scuba diver and not only someone with autism."

Other divers with autism appreciate the fact that underwater, they don't have to socially interact, and stimulation levels are kept low. Says Alice Kay in *Autism Parenting Magazine*: "The pressure of the water felt good to [our son] and the elimination of noises and stimulations vanished for the time he was underwater. . . . Making water bubbles, swimming on his back, and gliding through the water with turtles and fish allowed our son to finally be at peace."[14]

There are other benefits too, says Dan Johnson, who runs the Scuba for Autism program at Loves Park Scuba in Loves Park, Illinois, out of an eleven-foot pool. He points to one of his students, Liam, who has experienced incredible milestones. According to Liam's mother, Rebecah, "His speech has increased in ways, like even when he speaks to me, I'm blown away. For a spectrum mom with a child that had basically no language last year, that is huge for us."[15]

Elliott says adaptive scuba can help people off the spectrum as well. "For people with health problems such as chronic pain, diving can help lessen their symptoms.[16] . . . We've discovered the forgiving, weightless wonder of the water column provides the perfect gravity-free environment for those who might otherwise struggle on land. Underwater, we're all equal."[17]

Diveheart offers a directory of instructors who work with children on the spectrum around the country. Check it out at https://diveheart.org/instructor-directory. Elliott says for those seeking vacation destinations, they usually refer spectrum families to dive operations that have experience working with individuals with autism and other disabilities, such as Rainbow Reef in Key Largo, Gold Coast Scuba in Fort Lauderdale, and Dive Paradise in Cozumel. "We may also send families to one of our teams around the world or to one of our partners that have adaptive buddies and instructors on staff or independent adaptive scuba instructors or buddies that we've trained," he explains.

Here are some other autism-friendly dive centers where your child can receive scuba training. Not all programs may be as extensive as Diveheart's, so inquire ahead as to the extent of their training for special needs students. Some were, at least at one time, Certified Autism Centers, and may run tours, but resorts already described in this book, such as the Beaches properties and Ramon's Family Resort in Belize (discussed in chapters 7 and 12), will also offer opportunities to practice what children may have learned elsewhere.

- Florida
  - **Bright Horizons Diving**
    425 Page Bacon Road, Mary Esther, FL 32569
    850-716-0867
    https://www.brighthorizonsdiving.org

  - **Scubagym at Scuba Quest Orlando**
    8092 S. Orange Blossom Trail, Orlando, FL 32809
    407-851-6677
    https://www.thescubagym.com/index.php/special-needs

- Georgia
  - **Dive Georgia**
    168 Towne Lake Parkway, Woodstock, GA 30188
    404-285-8600
    http://www.divegeorgia.com

- Illinois
  - **Diveheart**
    900 Ogden Avenue #274, Downers Grove, IL 60515
    630-964-1983
    https://www.diveheart.org

- ○ **Loves Park Scuba**
  7307 N. Alpine Rd, Loves Park, IL 61111
  815-633-6969
  https://lovesparkscuba.com

- New York
  - ○ **Gotham Divers**
    125 E. 4th St, New York, NY 10003
    212-780-0879
    http://www.gothamdivers.com

- Virginia
  - ○ **BluWater Scuba**
    50 Culpeper Street, Warrenton, VA 20186
    571-379-7000
    https://bluwaterscuba.com

If your child loves water but isn't ready for scuba, how about paddle-boarding? This company gives lessons:

- **Riverbound Sports Paddle Co.**
  1425 E. University Drive, Suite B-102, Tempe, AZ 85281
  480-463-6686
  https://www.riverboundsports.com

BUILDING A VACATION AROUND SPECIFIC INTERESTS

> *"We recently took a trip to Wisconsin to visit a water park. My husband happened to stumble across the National Mustard Museum, which was close by, and we were able to get our son to visit it because mustard is Ayden's favorite food. He loved going and tells everyone about how we went especially for him. He loves the memory."*
> – Adventures-in-the-Makeing, parent, Novi, Michigan

It's been estimated that 75 percent to 95 percent of children with autism have intense or "circumscribed" interests that can occupy up to twenty-six hours of their time each week.[18] "Sometimes you hear this phrase, 'To meet the child where the child is,'" says John Gabrieli, a neuroscientist with the Massachusetts Institute of Technology. "If this is their natural motivating capacity, then rather than try to suppress it, it might be more helpful to the child to build on it."

In other words, instead of dragging a child where *you* want to go, why not build a trip around their specific interests? It stands to reason that a vacation revolving around a child's passion may help mitigate upset over any changes in routine as well as other hiccups that may develop along the way.

"One thing . . . common with kids on the autism spectrum is a love or obsession with trains and water," says Jennifer Hardy of Cruise Planners in Kent, Washington:

> With Alaska cruises, I help set my clients up on shore excursions and self-guided activities that take them to special train museums, railway experiences, salmon ladders, gold panning, and more. These are regular activities that anyone can reserve, but I know that they come with more sensory appropriate environments while also appealing to their specific interests.
>
> I have also planned a completely customized, partially escorted tour of New Zealand. [While] natural landscapes are the primary draw for most New Zealand tours, I was able to pull together an itinerary with significant focus on the Lord of the Rings folklore sites, trains, and hot springs, all of which were of pointed interest to my clients. We also ensured that each component of their experience would be in a small group of no more than ten guests, due to my client's discomfort around large groups.
>
> The added challenge with this itinerary: one family member was on the spectrum while another was traveling with a wheelchair. The level of planning and detail to satisfy both needs required something completely out of the realm of possibility for my clients to put together on their own. I partnered with my personal contact at the tour operator's office to completely customize every little component of this experience [and] make it exactly what my client desired.

Lisa Bertuccio of Vacation Your World in Otisville, New York, says she recently planned a trip for a family whose child with ASD lived and breathed all things trains and boats:

> The family said they had a wedding they were attending in Savannah, Georgia and wanted to make some sort of family vacation out of it, but they were concerned because their child was terrified of planes and did

not do too well on long car rides, so they were looking to take many breaks on the drive down.

I suggested that we could make the trip a train and boat theme since their son found comfort with both, perhaps take Amtrak since we could book a room on the train and their son could have a space away from any uncomfortable sounds or extra stimuli caused by being around a lot of people. I also mentioned visiting the train museum, the maritime museum, staying at a hotel near the harbor, and scheduling a riverboat ride. The idea of taking a train there had never even crossed their minds because they were so concerned with the typical ways of getting to their destination, i.e., car and plane. Sometimes the most obvious solution is the one that we overlook because of worry.

Passions can vary from typical to quirky, and while it isn't difficult to cater to an interest in art (museums); animals (zoos; wildlife parks; animal shelters; shows for dogs, cats, reptiles, and exotics); coins (mints as well as coin and money museums); stamps (museums, exhibitions); music (concerts, music halls of fame); and books (libraries, used and antique/rare bookstores), here are some suggestions of venues that might correspond with more unique interests. Preview them online to get a sense of whether the venue will interest your child. That peek may spark some advance excitement as you make that venue the centerpiece of an upcoming holiday.

Please note that for the sake of brevity, I've concentrated mostly on U.S. locations (listed alphabetically by state and then by city). This is a non-comprehensive list; further research on your part may reveal additional opportunities. Feel free to email me with your findings, and any passions I might have overlooked, at dawnbarclayauthor@gmail.com. These venues may or may not offer support for special needs. Call or check their websites to see if ASD modifications are available. Also verify addresses and dates/hours of operation. Some museums only open on certain days of the week; others are only open during certain months of the year; and still others require reservations. Some were temporarily closed at time of writing due to COVID but intended to reopen at a future date.

If you plan to visit a number of museums throughout the country, it might be worth joining the museum closest to you, especially if they are a member of the North American Reciprocal Museum (NARM) Association. Their connected membership opportunities might save you some money on entrance fees in your travels. Check https://narmassociation.org/how-it-works for more information.

Aviation (Planes)

There are a large number of aviation museums and air force museums in the United States. I've listed a few here as selected by BAA Training,[19] but you can find a more extensive list at https://www.airplanemuseums .com/list-of-air-museums-exhibits-airparks.htm. Also, as many parents with children on the spectrum know, plane spotting is one of the most popular activities for those with ASD. The Plane Finder Live Flight Tracker app gives live data on the planes you might see or hear overhead, including their make, age, departure city, and destination. During slow times on vacation, this could invigorate a child with ASD's interest and help him or her avoid boredom. Check it out at https://planefinder.net/apps.

- Arizona
  - **The Pima Air and Space Museum**
    6000 E. Valencia Rd., Tucson, AZ 85756
    520-574-0462
    https://pimaair.org

- California
  - **San Diego Air & Space Museum**
    2001 Pan American Plaza, San Diego, CA 92101
    619-234-8291
    https://sandiegoairandspace.org

- District of Columbia
  - **Smithsonian's National Air and Space Museum** (two locations)
    655 Jefferson Drive SW, Washington, DC 20560
    202-633-2214
    Steven F. Udvar-Hazy Center:
    14390 Air and Space Museum Pkwy, Chantilly, VA 20151
    703-572-4118
    https://airandspace.si.edu

- Kansas
  - **Kansas Aviation Museum**
    3350 S. George Washington Blvd, Wichita, KS 67210
    316-683-9242
    https://kansasaviationmuseum.org

- Ohio
  - **National Museum of the U.S. Air Force**
    1100 Spaatz Street, Wright-Patterson AFB (near Dayton), OH 45433
    937-255-3286
    https://www.nationalmuseum.af.mil

- New York
  - **The Intrepid Sea, Air, and Space Museum**
    Pier 86, W. 46th Street, New York, NY 10036
    212-245-0072
    https://www.intrepidmuseum.org

Clocks and Watches (Horological Museums)

- Connecticut
  - **American Clock & Watch Museum**
    100 Maple Street, Bristol, CT 06010
    860-583-6070
    https://www.clockandwatchmuseum.org

- Iowa
  - **Bily Clocks Museum**
    323 S. Main Street, Spillville, IA 52168
    563-562-3569
    https://www.bilyclocks.org

- Massachusetts
  - **The Willard House and Clock Museum**
    11 Willard Street, North Grafton, MA 01536
    508-839-3500
    https://willardhouse.org

- Pennsylvania
  - **National Watch and Clock Museum**
    514 Poplar Street, Columbia, PA 17512
    717-684-8261
    https://www.nawcc.org/visit

Computers and Videogames

- California
  - ○ **Computer History Museum**
    1401 N Shoreline Blvd, Mountain View, CA 94043
    650-810-1010
    https://computerhistory.org

  - ○ **The Tech Interactive**
    201 S. Market St, San Jose, CA 95113
    408-294-8324
    https://www.thetech.org

  - ○ **Intel Museum**
    2200 Mission College Blvd, Santa Clara, CA 95054
    408-765-5050
    https://www.intel.com/content/www/us/en/history/museum
    -visiting-intel.html

- District of Columbia
  - ○ **Smithsonian National Museum of American History**
    1300 Constitution Ave. NW, Washington, DC 20560
    202-633-1000
    https://americanhistory.si.edu/exhibitions/my-computing-device

- Georgia
  - ○ **Museum of Technology at Middle Georgia State University**
    100 University Pkwy, Library Building, Macon, GA 31206
    478-471-2801
    https://www.mga.edu/computing/technology-museum
      (Note: Collections may go on tour; call ahead before visiting.)

  - ○ **Computer Museum of America**
    5000 Commerce Pkwy, Roswell, GA 30076
    770-695-0651
    https://www.computermuseumofamerica.org

- Maryland
  - ○ **The System Source Computer Museum**
    338 Clubhouse Rd, Hunt Valley, MD 21031
    410-771-5544
    https://museum.syssrc.com

- Montana
  - **American Computer & Robotics Museum**
    2023 Stadium Dr #1A, Bozeman, MT 59715
    406-582-1288
    https://acrmuseum.org

- New York
  - **The Strong International Center for the History of Electronic Games**
    One Manhattan Square, Rochester, NY 14607
    585-263-2700
    https://www.museumofplay.org/about/icheg

- Pennsylvania
  - **Kennett Classic Computer Museum, Gifts and Hobby Shop**
    126 S. Union St, Kennett Square, PA 19348
    484-732-7041
    https://www.kennettclassic.com

  - **Large Scale Systems Museum**
    924 4th Ave, New Kensington, PA 15068
    No phone number available, but email info@lssmuseum.org
    https://www.mact.io

- Rhode Island
  - **Rhode Island Computer Museum**
    1130 Ten Rod Rd, Building C, Suite 103, North Kingstown, RI 02852
    401-741-6997
    https://www.ricomputermuseum.org

- Texas
  - **U.S. National Videogame Museum**
    8004 Dallas Pkwy, Frisco, TX 75034
    972-668-8400
    http://www.nvmusa.org

- Washington
  - **Microsoft Visitor Center**
    15010 NE 36th St, Building 92, Redmond, WA 98052
    425-703-6214
    https://www.microsoft.com/en-us/visitorcenter

- **Living Computers: Museum & Labs**
  2245 1st Ave S, Seattle, WA 98134
  206-342-2020
  https://www.livingcomputers.org

Construction

- District of Columbia
  - **National Building Museum**
    401 F St NW, Washington, DC 20001
    202-272-2448
    https://www.nbm.org

- Ohio
  - **The National Construction Equipment Museum**
    16623 Liberty Hi Rd, Bowling Green, OH 43402
    419-352-5616
    https://hcea.net

Dinosaurs

Note: In addition to these museums, many local science museums feature dinosaur exhibits. A listing of science museums follows later in this chapter.

- Colorado
  - **Denver Museum of Nature & Science**
    2001 Colorado Blvd, Denver, CO 80205
    303-370-6000
    https://www.dmns.org/visit/exhibitions/prehistoric-journey

- District of Columbia
  - **The Smithsonian National Museum of Natural History**
    1000 Madison Drive NW, Washington, DC 20560
    202-633-1000
    https://naturalhistory.si.edu/exhibits/david-h-koch-hall-fossils-deep-time

- Illinois
  - ○ **The Field Museum**
    1400 S. DuSable Lake Shore Dr, Chicago, IL 60605
    312-922-9410
    https://www.fieldmuseum.org/exhibitions/evolving-planet

- Massachusetts
  - ○ **Museum of Science**
    1 Science Park, Boston, MA 02114
    617-723-2500
    https://www.mos.org/exhibits/triceratops-cliff

- New Jersey
  - ○ **Liberty Science Center**
    222 Jersey City Blvd, Jersey City, NJ 07305
    201-200-1000
    https://lsc.org/explore/exhibitions

- New York
  - ○ **American Museum of Natural History**
    200 Central Park West, New York, NY 10024
    212-769-5100
    https://www.amnh.org/exhibitions/permanent/orientation
    -center/the-titanosaur
    https://www.amnh.org/exhibitions/permanent/saurischian
    -dinosaurs/tyrannosaurus-rex

- North Carolina
  - ○ **North Carolina Museum of Natural Sciences**
    11 W. Jones St, Raleigh, NC 27601
    919-707-9800
    https://naturalsciences.org/exhibits/permanent-exhibits/nature
    -exploration-center

- North Dakota
  - ○ **Badlands Dinosaur Museum**
    188 Museum Dr East, Dickinson, ND 58601
    701-456-6225
    http://dickinsonmuseumcenter.com
    (Note: Seven other museums are nearby; see https://www.nd
    tourism.com/best-places/8-stops-north-dakota-dinosaur-tour.)

- Texas
  - **Perot Museum of Nature and Science**
    2201 N. Field St, Dallas, TX 75201
    214-428-5555
    https://www.perotmuseum.org/exhibits-and-films/permanent
    -exhibit-halls/life-then-and-now-hall.html

- Wyoming
  - **Wyoming State Museum**
    2301 Central Ave, Cheyenne, WY 82001
    307-777-7022
    https://wyomuseum.wyo.gov/index.php/current-exhibits-list
    /79-r-i-p-rex-in-pieces

  - **University of Wyoming Geological Museum**
    200 N 9th St (corner of 9th and Lewis), Laramie, WY 82072
    307-766-2646
    http://www.uwyo.edu/geomuseum/exhibits

  - **Wyoming Dinosaur Center**
    110 Carter Ranch Rd, Thermopolis, WY 82443
    307-864-2997
    https://wyomingdinosaurcenter.org

Elevators

- Illinois
  - **J.H. Hawes (Grain) Elevator Museum**
    199-293 SW 2nd Street, Atlanta, IL 61723
    309-830-8306
    No website available, informational article at https://www.ilfb
    partners.com/family/hawes-grain-elevator-museum-in-atlanta/

- Massachusetts
  - **The Elevator Museum Inc.**
    145 Essex St, Haverhill, MA 01832
    603-828-1849
    https://theelevatormuseuminc.org

- Virginia
  - **elevaTOURS Elevator Museum**
    909 4th Street SE, Roanoke, VA 24013
    No phone number available; you must make an appointment to visit. See the form on the website at https://www.dieselducy .com/elevator-museum.html.
    (The museum is inside a storage facility and is run by someone with ASD; many elevator videos are offered on the site.)

- Outside the United States/Canada
  - **Canadian Grain Elevator Discovery Centre**
    2119 19th Ave, Nanton, AB T0L 1R0, Canada
    403-646-1146
    https://www.nantongrainelevators.com

Insects

Be sure to also contact your local zoo for pertinent exhibits.

- Arizona
  - **Butterfly Wonderland**
    9500 E. Via de Ventura, Scottsdale, AZ 85256
    480-800-3000
    https://butterflywonderland.com

- Colorado
  - **May Natural History Museum**
    710 Rock Creek Canyon Road, Colorado Springs, CO 80926
    719-576-0450
    https://coloradospringsbugmuseum.com

- District of Columbia
  - **O. Orkin Insect Zoo, Smithsonian Institution/Museum of Natural History**
    1000 Madison Drive NW, Washington, DC 20560
    202-633-1000
    https://naturalhistory.si.edu/exhibits/o-orkin-insect-zoo

- Louisiana
  - **Insectarium at Audubon Nature Institute**
    504-524-2847
    https://audubonnatureinstitute.org/insectarium
    At time of writing, the Audubon Butterfly Garden and Insectarium was in the process of moving from the U.S. Custom House on Canal Street to Audubon Aquarium of the Americas. Contact for location and details.

- Missouri
  - **Bayer Insectarium at Saint Louis Zoo**
    1 Government Dr, St. Louis, MO 63110
    314-781-0900
    https://www.stlzoo.org/visit/thingstoseeanddo/discovery corner/insectarium

- Pennsylvania
  - **Philadelphia Insectarium and Butterfly Pavilion**
    8046 Frankford Ave, Philadelphia, PA 19136
    215-335-9500
    https://www.phillybutterflypavilion.com

- Texas
  - **The Houston Museum of Natural Science**
    5555 Hermann Park Dr, Houston, TX 77030
    713-639-4629
    http://www.hmns.org

  - **The Bug and Reptile Museum**
    1118 Charleston Beach Rd W, Bremerton, WA 98312
    No phone number available; email help@bugmuseum.com
    http://www.bugmuseum.com

  - **Pacific Science Center**
    200 Second Avenue North, Seattle, WA 98109
    206-443-2001
    https://www.pacificsciencecenter.org/exhibits/living-exhibits

Rocks

You can find a directory of mineral and geology museums around the United States, including the Alaska Museum of Science and Nature in Anchorage and the American Museum of Natural History in New York City, at https://www.mindat.org/museums. Visit VirtualMuseumofGeology .com for a directory of rock shops and geology attractions throughout the United States and Canada. If your child would like to see rock extractions in real time, why not visit a quarry? Many have observation decks and visiting hours. Members of the National Building Granite Quarries Association can be found at http://nbgqa.com/quarries, but may not necessarily offer tours. Here are a few quarries that do:

- Illinois
  - **Thornton Quarry**
    114 N. Hunter Street, Thornton, IL 60476
    708-877-6589
    https://www.thorntonilhistory.com

- North Carolina
  - **North Carolina Granite Corporation**
    151 Granite Quarry Trail, Mt Airy, NC 27030
    336-786-5141
    https://nbgqa.com/nc-granite-corp/

- Texas
  - **Alibates Flint Quarries Tours**
    P.O. Box 1460, Fritch, TX 79036
    806-857-3151
    http://www.ohranger.com/alibates-flint-quarries/planning-your-visit

- Vermont
  - **Rock of Ages Visitors Center**
    558 Graniteville Road, Graniteville, VT 05654
    802-476-3119
    https://rockofages.com/tourism/

SCIENCE

If your child dreams of learning more about space, how gadgets function, how ecosystems work, the discovery of electricity, and so much more, why not visit a science museum on an upcoming trip? For a more complete list, check out the more than five hundred members of the Association of Science and Technology Centers at https://www.astc.org/find-a-science-center. The following are considered some of the best in the country:

- Arkansas
  - **Museum of Discovery**
    500 President Clinton Ave, Suite 150, Little Rock, AR 72201
    501-396-7050
    https://museumofdiscovery.org

- California
  - **California Science Center**
    700 Exposition Park Dr, Los Angeles, CA 90037
    323-724-3623
    https://californiasciencecenter.org

  - **Exploratorium**
    Pier 15, Embarcadero at Green St., San Francisco, CA 94111
    415-528-4444
    https://www.exploratorium.edu

- District of Columbia
  - **National Air and Space Museum** (two locations)
    655 Jefferson Drive SW, Washington, DC 20560
    202-633-2214
    Steven F. Udvar-Hazy Center
    14390 Air and Space Museum Pkwy, Chantilly, VA 20151
    703-572-4118
    https://airandspace.si.edu

- Illinois
  - **Museum of Science and Industry**
    5700 S. DuSable Lake Shore Dr, Chicago, IL 60637
    773-684-1414
    https://www.msichicago.org

- Massachusetts
  - **Museum of Science**
    1 Science Park, Boston, MA 02114
    617-723-2500
    https://www.mos.org

- Minnesota
  - **Science Museum of Minnesota**
    120 W. Kellogg Blvd, St. Paul, MN 55102
    651-221-9444
    https://new.smm.org

- New Jersey
  - **Liberty Science Center**
    222 Jersey City Blvd, Jersey City, NJ 07305
    201-200-1000
    https://lsc.org

- North Carolina
  - **Discovery Place**
    301 N. Tryon St, Charlotte, NC 28202
    704-372-6261
    https://www.discoveryplace.org

- Ohio
  - **COSI: The Center of Science and Industry**
    333 W. Broad St, Columbus, OH 43215
    614-228-2674
    https://www.cosi.org

- Pennsylvania
  - **Franklin Institute**
    222 N. 20th St, Philadelphia, PA 19103
    215-448-1200
    https://www.fi.edu

  - **The Academy of Natural Sciences of Drexel University**
    1900 Benjamin Franklin Pkwy, Philadelphia, PA 19103
    215-299-1000
    https://ansp.org

Transportation—Railroads and Trains, Trucks, and Cars

TRANSIT MUSEUMS/GENERAL

Six hundred transit museums are listed at https://en.wikipedia.org/wiki/List_of_transport_museums.

- **Railway/trains**
  - California
    - **California State Railroad Museum**
      111 I St, Sacramento, CA 95814
      916-323-9280
      https://www.californiarailroad.museum

  - Colorado
    - **Colorado Railroad Museum**
      17155 W. 44th Ave, Golden, CO 80403
      303-279-4591
      https://coloradorailroadmuseum.org

  - Illinois
    - **Illinois Railway Museum**
      7000 Olson Rd, Union, IL 60180
      815-923-4391
      https://www.irm.org

  - Maryland
    - **B&O Railroad Museum**
      901 W. Pratt St, Baltimore, MD 21223
      410-752-2490
      http://www.borail.org

  - Massachusetts
    - **Berkshire Scenic Railway Museum**
      10 Willow Creek Rd, Lenox, MA 01240
      413-637-2210
      https://www.berkshiretrains.org

○ Nevada
  ▪ **Nevada Northern Railway Museum**
    1100 Ave A, Ely, NV 89301
    775-289-2085
    https://www.nnry.com

○ New York
  ▪ **NY Transit Museum**
    99 Schermerhorn St, Brooklyn, NY 11201
    718-694-1600
    https://www.nytransitmuseum.org

○ North Carolina
  ▪ **Smoky Mountain Trains Museum**
    100 Greenlee St, Bryson City, NC 28713
    800-872-4681
    https://www.gsmr.com/smoky-mountain-trains-museum

• **Trucks and tanks**
  ○ California
    ▪ **Los Angeles County Fire Museum**
      16400 Bellflower Blvd, Bellflower, CA 90706
      562-925-0234
      https://www.lacountyfiremuseum.com

  ○ Iowa
    ▪ **Iowa 80 Trucking Museum**
      505 Sterling Drive, Walcott, IA 52773
      563-468-5500
      https://iowa80truckingmuseum.com

  ○ New Jersey
    ▪ **Military Technology Museum of New Jersey**
      InfoAge, Building 9011, 2201 Marconi Road, Wall Township, NJ 07719
      848-404-9774
      https://mtmnj.org

- Pennsylvania
  - **Mack Trucks Historical Museum**
    2402 Lehigh Pkwy S, Allentown, PA 18103
    610-351-8999
    https://macktruckshistoricalmuseum.org

- Wyoming
  - **National Museum of Military Vehicles**
    6419 US Highway 26, Dubois, WY 82513
    307-455-3802
    https://www.nmmv.org

- **Cars**
  - Alabama
    - **Barber Vintage Motorsports Museum**
      6030 Barber Motorsports Pkwy, Birmingham, AL 35094
      205-699-7275
      https://www.barbermuseum.org

  - California
    - **Blackhawk Museum**
      3700 Blackhawk Plaza Circle, Danville, CA 94506
      925-736-2280
      https://blackhawkmuseum.org

    - **Petersen Auto Museum**
      6060 Wilshire Blvd, Los Angeles, CA 90036
      323-930-2277
      https://www.petersen.org

    - **California Auto Museum**
      2200 Front St, Sacramento, CA 95818
      916-442-6802
      https://www.calautomuseum.org

  - Florida
    - **Don Garlits Museum of Drag Racing**
      13700 SW 16th Ave, Ocala, FL 34473
      352-245-8661
      https://garlits.com

○ Illinois
- **Volo Museum**
  27582 Volo Village Rd, Volo, IL 60073
  815-385-3644
  https://www.volocars.com

○ Indiana
- **National Automobile and Truck Museum**
  1000 Gordon M Buehrig Pl, Auburn, IN 46706
  260-925-9100
  https://natmus.org

- **Indianapolis Motor Speedway Museum**
  4750 W. 16th St, Indianapolis, IN 46222
  317-492-6784
  https://imsmuseum.org

○ Massachusetts
- **Larz Anderson Auto Museum**
  15 Newton St, Brookline, MA 02445
  617-522-6547
  https://larzanderson.org

○ Michigan
- **Henry Ford Museum of American Innovation**
  20900 Oakwood Blvd, Dearborn, MI 48124
  313-982-6001
  https://www.thehenryford.org

○ Nevada
- **National Automobile Museum**
  1 Museum Drive, Reno, NV 89501
  775-333-9300
  https://automuseum.org

○ Pennsylvania
- **Simeone Foundation Automobile Museum**
  6825 Norwitch Dr, Philadelphia, PA 19153
  215-365-7233
  https://simeonemuseum.org

- ○ Tennessee
  - **Lane Motor Museum**
    702 Murfreesboro Pike, Nashville, TN 37210
    615-742-7445
    https://www.lanemotormuseum.org

- ○ Washington
  - **LeMay-America's Car Museum**
    2702 East "D" St, Tacoma, WA 98421
    253-779-8490
    https://www.americascarmuseum.org

- **Trolleys and Streetcars**
  Find additional streetcar museums and resources at https://www
  .dctrolley.org/otherMuseums.html.

  - ○ California
    - **Southern California Railway Museum**
      2201 S. "A" Street, Perris CA 92570
      951-943-3020
      https://socalrailway.org

  - ○ Connecticut
    - **Connecticut Trolley Museum**
      58 North Rd, East Windsor, CT 06088
      860-627-6540
      https://www.ct-trolley.org

  - ○ Maryland
    - **Baltimore Streetcar Museum**
      1901 Falls Road, Baltimore, MD 21211
      410-547-0264
      https://www.baltimorestreetcarmuseum.org

  - ○ Pennsylvania
    - **Electric City Trolley Museum**
      300 Cliff Street, Scranton, PA 18503
      570-963-6590
      http://www.ectma.org

Car and truck afficionados might also enjoy the following events, in addition to the museums.

- **Racetracks and racing events** (remember your noise-canceling headphones!)
  - California
    - **WeatherTech Raceway Laguna Seca**
      1021 Monterey Salinas Hwy, Salinas, CA 93908
      831-242-8201
      https://www.co.monterey.ca.us/government/government
      -links/weathertech-raceway

    - **Sonoma Raceway**
      29355 Arnold Dr, Sonoma, CA 95476
      800-870-7223
      https://www.sonomaraceway.com

  - Connecticut
    - **Lime Rock Park**
      60 White Hollow Rd, Lakeville, CT 06039
      860-435-5000
      https://limerock.com

  - Florida
    - **Daytona International Speedway**
      1801 W. International Speedway Blvd, Daytona Beach, FL 32114
      800-748-7467
      https://www.daytonainternationalspeedway.com

  - Indiana
    - **Indianapolis Motor Speedway**
      4790 W. 16th St, Indianapolis, IN 46222
      317-492-8500
      https://www.indianapolismotorspeedway.com

  - New York
    - **Watkins Glen International Raceway**
      2790 County Route 16, Watkins Glen, NY 14891
      607-535-2486
      https://www.theglen.com

- ○ Texas
  - • **Circuit of the Americas**
    9201 Circuit of the Americas Blvd, Austin, TX 78617
    512-301-6600
    http://circuitoftheamericas.com

- ○ Utah
  - • **Utah Motorsports Campus**
    512 Sheep Ln, Grantsville, UT 84074
    435-277-8000
    https://www.utahmotorsportscampus.com

- ○ Virginia
  - • **Virginia International Raceway**
    1245 Pine Tree Rd, Alton, VA 24520
    434-822-7700
    https://virnow.com

- ○ Wisconsin
  - • **Road America**
    N7390 State Highway 67, Plymouth, WI 53073
    920-892-4576
    https://www.roadamerica.com

Looking for more?

- NASCAR racing events: https://www.nascar.com/nascar-cup-series/2021/schedule
- American Flat Track (motorcycle racing) events: https://www.americanflattrack.com/events
- AMA Pro Racing events: http://www.amaprohillclimb.com/events/default
- Motocross Racing events: https://mxsportsproracing.com/events
- ATV MX Racing events: https://atvmotocross.com/events
- Monster truck events: https://www.monsterjam.com/en-US/tickets

Woodworking

If your child yearns to create objects out of wood, these museums will provide the inspiration.

- Connecticut
  - **Yale University Art Gallery**
    1111 Chapel St, New Haven, CT 06510
    203-432-0601
    https://artgallery.yale.edu

- District of Columbia
  - **Renwick Gallery of the Smithsonian American Art Museum**
    1661 Pennsylvania Avenue NW, Washington, DC 20006
    202-633-7970
    https://www.si.edu/museums/renwick-gallery

- Georgia
  - **High Museum**
    1280 Peachtree St NE, Atlanta, GA 30309
    404-733-4400
    https://high.org

- Massachusetts
  - **Museum of Fine Arts Boston**
    465 Huntington Avenue, Boston, MA 02115
    617-267-9300
    https://www.mfa.org

- Minnesota
  - **Minneapolis Institute of Arts**
    2400 Third Ave S, Minneapolis, MN 55404
    612-870-3000
    https://new.artsmia.org

  - **AAW Gallery of Wood Art**
    222 Landmark Center, 75 5th St W, St Paul, MN 55102
    651-484-9094
    http://galleryofwoodart.org

- New York
  - **Antique Boat Museum**
    750 Mary St, Clayton, NY 13624
    315-686-4104
    https://www.abm.org

- **Hudson Valley Maritime Museum (Wooden Boat School)**
  50 Rondout Landing, Kingston, NY 12401
  845-338-0071
  https://www.hrmm.org/woodworking—craft-classes.html
    The Hudson Valley Maritime Museum also offers classes.

- **The Shaker Museum**
  202 Shaker Road, New Lebanon, NY 12125
  518-794-9100
  https://www.shakermuseum.us/tag/woodworking

- North Carolina
  - **Grovewood Gallery and Museum**
    111 Grovewood Rd, Suite 2, Asheville, NC 28804
    828-253-7651
    https://www.grovewood.com

  - **Mint Museum**
    2730 Randolph Road, Charlotte, NC 28207
    704-337-2000
    https://mintmuseum.org

- Pennsylvania
  - **Wharton Esherick Museum**
    1520 Horseshoe Trail, Malvern, PA 19355
    610-644-5822
    https://whartonesherickmuseum.org

  - **Center for Art in Wood**
    141 N. 3rd Street, Philadelphia, PA 19106
    (215) 923-8000
    https://centerforartinwood.org

- Wisconsin
  - **Racine Art Museum**
    441 Main Street, Racine, WI 53403
    262-638-8300
    https://www.ramart.org

- Outside the United States/Canada
  - **MacLachlan Woodworking Museum**
    2993 Hwy 2 East, Kingston, ON K7L 4V1, Canada
    613-542-0543
    https://www.woodworkingmuseum.ca

Quirky

You can find over seven hundred museums of unique interest at Atlas Obscura. See https://www.atlasobscura.com/things-to-do/united-states /museums-and-collections. Here are just a few to whet your appetite.

- District of Columbia
  - **International Spy Museum**
    700 L'Enfant Plaza SW, Washington, DC 20024
    202-393-7798
    https://www.spymuseum.org

- Idaho
  - **Idaho Potato Museum**
    130 NW Main St, Blackfoot, ID 83221
    208-785-2517
    https://idahopotatomuseum.com

- Kansas
  - **The Kansas Barbed Wire Museum**
    120 W. 1st St, La Crosse, KS 67548
    785-222-9900
    http://www.rushcounty.org/BarbedWireMuseum

- Louisiana
  - **New Orleans Pharmacy Museum** (early medications, super-stitions, cures)
    514 Chartres St, New Orleans, LA 70130
    504-565-8027
    https://www.pharmacymuseum.org

- Maine
  - **International Cryptozoology Museum** (Bigfoot, Loch Ness Monster, etc.)
    32 Resurgam Pl, Portland, ME 04102
    No phone number available
    https://cryptozoologymuseum.com

- Minnesota
  - **Spam Museum**
    101 3rd Ave NE, Austin, MN 55912
    507-437-5100
    https://www.spam.com/museum

- New Mexico
  - **International UFO Museum and Research Center**
    14 N. Main St, Roswell, NM 88203
    575-625-9495
    https://www.roswellufomuseum.com

- Pennsylvania
  - **Mutter Museum at the College of Physicians of Philadelphia** (body parts, including skulls, and medical equipment)
    19 S. 22nd St, Philadelphia, PA 19103
    215-563-3737
    http://muttermuseum.org

- Tennessee
  - **Salt and Pepper Shaker Museum**
    461 Brookside Village Way, Gatlinburg, TN 37738
    865-430-5515
    https://thesaltandpeppershakermuseum.com

- Texas
  - **National Museum of Funeral History**
    415 Barren Springs Dr, Houston, TX 77090
    281-876-3063
    https://www.nmfh.org

- Wisconsin
  - **National Mustard Museum**
    7477 Hubbard Ave, Middleton, WI 53562
    608-831-2222
    https://mustardmuseum.com

CHAPTER 17 ACTION STEPS

1. Consider visiting cities that have devoted themselves to autism friendliness, like Mesa, Arizona, and Myrtle Beach, South Carolina.
2. Don't be afraid to venture beyond theme parks and autism-friendly cruises, but always remember that unique travel still requires safety precautions.
3. If a city has enough autism-friendly attractions, combine them into an itinerary and create your own autism-friendly vacation, using strategies from other chapters of this book.
4. Non-team sports like golf, skiing, and scuba diving have programs adapted for neurodiverse children. Why not include them in a holiday or build a vacation around them?
5. Kids with ASD often have unique interests that have become passions. Building a vacation around museums and other venues catering to those interests can create memories to last a lifetime.

Part VI

RESOURCES AND ASSISTANCE

18

RESOURCES

MENTAL HEALTH ORGANIZATIONS

There isn't enough space to list all the resources and blogs that abound concerning autism spectrum disorder (ASD) and other mental health issues, but if you're looking for more information, here are just a few of the organizations that may help. Unless otherwise indicated, the phone numbers given are generally customer service numbers, not emergency hotlines.

In General

- **American Psychological Association (APA):** 800-374-2721; https://www.apa.org
- **American Psychiatric Association:** 888-357-7924; https://www.psychiatry.org
- **Mental Health America (MHA):** 800-969-6642; https://mhanational.org
- **National Alliance on Mental Illness (NAMI):** 703-524-7600; https://www.nami.org
- **National Institute of Mental Health (NIMH):** 866-615-6464; https://www.nimh.nih.gov
- **Substance Abuse and Mental Health Services Administration (SAMHSA):** 800-662-4357; https://www.samhsa.gov

Autism

- **American Academy of Pediatrics (AAP):** News and journals: 888-227-1770; https://www.aappublications.org/search/autism

- **American Autism Association:** https://www.myautism.org, no phone number provided, but email is info@myautism.org.
- **Asperger/Autism Network (AANE):** 617-393-3824; https://www.aane.org
- **Association of University Centers on Disabilities (AUCD):** 301-588-8252; https://www.aucd.org
- **Autism NOW:** 855-828-8476; https://autismnow.org
- **Autism Science Foundation (ASF):** 914-810-9100; https://autismsciencefoundation.org
- **Autism Society:** 800-328-8476; https://www.autism-society.org
- **Autism Speaks:** 888-288-4762; 888-772-9050 (Espanol); https://www.autismspeaks.org
- **Autistic Self Advocacy Network (ASAN):** Phone number unavailable; email: info@autisticadvocacy.org; https://autisticadvocacy.org
- **Center for Autism and Related Disorders (CARD):** 877-448-4747; https://www.centerforautism.com
- **The Color of Autism:** 313-444-9035; https://thecolorofautism.org
- **Grupo Salto (in Spanish):** 773-757-9691; https://gruposalto.org
- **National Autism Association:** 877-622-2884; https://nationalautismassociation.org/
- **U.S. Autism Association:** 888-928-8476; https://www.usautism.org

Attention-Deficit/Hyperactivity Disorder

- **ADDitude:** 212-417-9700; https://www.additudemag.com
- **Attention Deficit Disorder Association (ADDA):** 800-939-1019; https://www.add.org
- **Centre for ADHD Awareness, Canada:** 800-807-0090; https://caddac.ca
- **Children and Adults with Attention-Deficit/Hyperactivity Disorder (CHADD):** 301-306-7070; https://chadd.org
- **LD Online:** Phone number unavailable; email: ldonline@weta.org; http://www.ldonline.org

Bipolar Syndrome

- **Depression and Bipolar Support Alliance (DBSA):** 800-826-3632; https://www.dbsalliance.org
- **International Bipolar Foundation:** Suicide Crisis Line: 800-273-8255; https://ibpf.org

Tourette Syndrome

- **Tourette Association of America:** 888-486-8738; https://tourette.org
- **National Institute of Neurological Disorders and Stroke (NIH):** (800) 352-9424; https://www.ninds.nih.gov

Borderline Personality Disorder

- **National Education Alliance for Borderline Personality Disorder:** Phone number unavailable; email: info@neabpd.org; https://www.borderlinepersonalitydisorder.org
- **United Brain Association:** (202) 768-8000; https://unitedbrainassociation.org/brain-resources/borderline-personality-disorder

Travel Related

- **Family Travel Association:** 307-250-1981; email: info@familytravel.org; https://familytravel.org
- **Autism Travel (International Board of Credentialing and Continuing Education Standards [IBCCES]):** 877-717-6543; https://autismtravel.com
- **Society for Accessible Travel and Hospitality (SATH):** 212-447-7284; http://sath.org
- **TravelAbility:** 415-339-0578; https://travelability.net

TRAVEL ADVISORS TO HELP BOOK YOUR TRIP

Unless indicated otherwise, each of the travel advisors listed here has received their Certified Autism Travel Professional (CATP) designation.

While all might not be quoted in the book, these professionals took the time and effort to contribute to the information contained herein. They are listed alphabetically by last name as opposed to geographically because, thanks to the wonders of telephone and videoconferencing, you can work with them from anywhere to ensure your upcoming vacation will be as seamless and successful as possible.

Please note that longevity in the travel industry for these professionals was accurate as of the time of writing. You can find updated contact information for these travel counselors by searching the IBCCES website at https://bit.ly/2YIEfNl or https://apps.ibcces.org/cert/registry (enter "Certified Autism Travel Professional" under "Certification").

- **Dawn Baeszler is with AAA Travel** in Lawrenceville, New Jersey. She's been working in travel for three years, though she's traveled extensively with her children for ten. She pursued the CATP designation because

  > both my children are on the spectrum and have sensitivity and sound issues. They both attended a special education school for dyslexia. I have been an advocate for children with learning disabilities for fourteen years and have served as an advocate in schools for curriculum reform to offer strategies for disabilities associated with dyslexia, ADHD, autism, and executive functioning. I have testified to Congress for learning reforms in those areas as well. My son has severe dyslexia, which has inhibited his ability to read signs, which is a crucial issue for my child when traveling. . . . The CATP designation [informs my] clients that I am educated in the latest methods of travel [for those with] spectrum-based disabilities.

- **Leslie Berkman is with Berkie Travel** in Cranford, New Jersey. She's worked in travel for two decades. She got her CATP designation because Sandals/Beaches was offering a discount to travel agents to become CATP-certified. "The training gave me an understanding of the autism spectrum, and its various levels. I've renewed several times, so I have the appropriate knowledge to advise my clients accordingly," she explains.

- **Lisa Bertuccio of Vacation Your World** in Otisville, New York, a home-based agency, has spent the past two years in travel:

I was approached by a potential client a year ago to help plan the first trip the family would take together since learning of their son's autism diagnosis. Having never planned a vacation for a family with a child on the spectrum before, I knew [that this] would not be like planning vacations for families with neurotypical children. This led me to IBCCES's CATP program. I let the family know that I would need to complete this program first . . . and fortunately they understood and were appreciative that I would educate myself on the area of autism travel before jumping in to plan their vacation.

This designation has helped me . . . [learn to ask] questions that I would normally not ask, such as what sensory issues I needed to be aware of. [I discovered] what resources are available for children and parents, [how] planning [can] lessen the anxiety and stress of parents and children before and during travel, and . . . that although two children may be diagnosed with the same disorder, their needs can be very different. I can't assume that there is a one-size-fits-all way to plan for a family with a child who has ASD. [The training] has also helped me connect with a wonderful community of families who never thought it was possible to travel and explore the world with their children. They are now able to see that it is . . . not only possible, but that there are many things that can help ensure that their children's needs are taken care of before and during their trip.

- **Chandra Nims Brown of Cruise Planners—Global Travel Pros** in Williamsburg, Virginia, has worked in travel for twelve years. She studied for the CATP designation because "I have a very good friend whose son is autistic. We had several conversations about the importance of finding locations and vacation options that provide the services needed for her family. . . . My CATP designation has helped me connect with an underserved group in my community. All people, regardless of their particular situation, should be able to travel safely and enjoyably."

- **Betty J. Burger, ECC, MCC, ACC, of BGB Travel and Leisure, LLC,** in St. Marys, Georgia, planned vacations for more than ten years. During that time, she helped multi-generational families travel with members who had special physical needs. She chose to get her CATP designation because having friends with autistic children and knowing the difficulties they have in being able to vacation as a family, I wanted to learn more about the issues they face. . . . Special needs families are not "one size fits all" by any means. Each situation demands assessment of motor skills, social skills, communication skills, emotional status, and sensory needs. [Because of this training], I developed a better understanding and was able to approach special needs groups well prepared to . . . accurately assess the level of support needed and make appropriate recommendations for activities and resorts . . . that met their needs and where the entire family could enjoy a fun-filled vacation.

Betty recently retired from the travel industry, but her comments continue to remain relevant and insightful.

- **Sandra Chu is with The Enchanted Traveler** located in Puyallup, Washington, though she is based in Rockford, Illinois. She's worked in travel for almost five years:

    I specialize in family travel and if you extend any family far enough, you will find someone with some sort of special sensory needs. As part of my services, I provide customized itineraries that are designed to meet the individual needs of each family. Through the CATP training, I learned a lot about the needs of families whose members had sensory issues. To maintain the designation, we are required to study several hours per year of continuing education, which continues to expand my knowledge.

    Holding the designation enhances my credibility as a family travel planner and distinguishes me from other, more traditional travel agents. . . . We are all on a spectrum of sensory sensitivities. Those with autism are simply a little further down on the spectrum than others. Many people who might not be considered "autistic" do have some enhanced sensory issues. Some are overly sensitive to certain stimuli, while others may be under-sensitive. Working with families with autism involves understanding the sensory triggers of that person and finding the right solution to meet those challenges. It is not a "one-size-fits-all" approach.

- **Terence Concannon works with Go Lake Havasu** in Lake Havasu City, Arizona. He's worked in travel for twenty years and studied for his CATP designation because "we care about all of our residents and our visitors and want everyone who visits Lake Havasu to feel welcome, included, and cared for. The training has helped both me and my staff better understand the challenges faced by persons on the spectrum and their caregivers."

- **Markeisha Hall of Funtherapy Travel** in Rosena Ranch, California, was a former special education teacher and a go-to person for special needs travel even before she became an independent travel agent in 2016. Her CATP designation has "helped me by allowing [more] parents to let down their walls and explain their special vacation needs. Seeing the number of families who do not travel was eye-opening." She says that a pervasive misconception among families on the spectrum is that

  > travel will be very difficult, and they will not be able to enjoy themselves . . . but in some respects, it is the same [as vacationing with neurotypical children.] Traveling with children is a task in itself, trust me. We were the family on a cruise with a double stroller. But the main difference is preparing your kiddo for the new experience and making sure the destination is aware of what your needs will look like once you get there.

  Her advice for neurodiverse travelers: "Consider traveling with other families that have special needs. We are actually starting itineraries for this very purpose."

- **Jonathan Haraty of Adventure Luxury Travel** in Whitman, Massachusetts, has spent the last twenty-one years in travel. He obtained his CATP designation because

  > I have friends [with] children on the spectrum, and they would ask me for travel recommendations. When I saw [IBCCES] was offering a new travel agent certification/designation I signed up for the training because this travel niche was just starting to evolve. I already knew that Beaches Resorts and Royal Caribbean had been training their staff to work with [those] on the spectrum. So, the certification just seemed like a nice fit with what I was doing already.
  >
  > [I've learned] that there are many different variations on the autism spectrum, and [that] more and more attractions, cruises, hotels, and resorts are training their people to work with [this

population]. This gives me [many] more options when planning [their] vacations. [By training travel agents] how to talk to parents and let them know that it is really okay to take a vacation with their child with autism . . . [my certification] has given new clients more confidence in booking with me.

- **Jennifer Hardy is both the franchise owner and an Accessible Travel Specialist with Cruise Planners** in Kent, Washington, and the parent of four children, aged three to twelve. Two have been diagnosed with ASD, and the other two with diagnoses that are "similar in manifestation but have different root causes." She has been in travel for sixteen years, ten as a travel advisor, and began traveling with her children on the spectrum when they were ages one and two, well before they were diagnosed. She has worked with families on the spectrum as well as those diagnosed with fetal alcohol spectrum disorders, sensory processing disorder, developmental delays, anxiety disorders, and more:

> I pursued the CATP designation the moment I first heard about it. I had been helping families with children on the spectrum for years, but it was completely based upon my personal experience, and I had not received any formal training prior to 2019. The designation has credentialed the work I do for families, and personally it has helped me to be more confident in my ability to work with families who have different needs. It solidified a lot of what I was already doing and provided me with some new questions to help me better qualify my clients.
>
> The designation has also provided me with the credibility I needed to partner with various therapy centers and autism organizations. This allows me a better opportunity to interact with families and open their minds to the possibility of [taking] a family vacation—something most feel is completely out of reach. The most important takeaway from the CATP program was that every individual with autism is just that—an individual. No two people on the autism spectrum are the same, and every single person deserves to be treated and cared for with respect. [You must] approach each family without a preconception of what their needs will be, and instead discover this through conversation. You cannot learn about a person's needs by checking boxes off a list. Rather, you really need to get to know them as a complete individual: What do they like and dislike? What are

their favorite activities and interests? What has travel been like for them in the past? What kind of accommodations are in place at home and school? What are their potential triggers and what is it like for the family when those triggers occur?

While we have two children who are not on the spectrum, they are not neurotypical. Impacted by significant prenatal histories, our younger children struggle more with impulsivity and anxiety or sensory-related behaviors than our children on the spectrum. It is much harder to travel with our younger children than it is with our older autistic ones. The good news is that all the resources available for those on the spectrum are just as useful for our other children. We find that people are willing to extend accommodations when you explain how they are helpful. But because our younger children's behaviors are so much more extensive, we would not be able to travel at all if it weren't for the accommodations designed for autistic travelers.

- **Georgiann Jaworskyj, CTC, ACC of Custom Travel Services, Inc.**, in Merchantville, New Jersey, has spent forty-three years in travel. She says she was first introduced to the CATP program by Sandals (Beaches for families) Resorts and realized how valuable it could be for families with children diagnosed on the spectrum. "[The designation] educated me on the subtle issues the families experience. . . . I [now] better realize just how different each child on the spectrum can be and how their needs have to be dealt with individually. . . . I feel . . . I am better prepared [to work with those on the spectrum] than agents who have not received this training."

- **Carl W. Johnson, president of Adventures in Golf** (incorporated as Hamill Travel Services, Inc.) in Amherst, New Hampshire, has spent more than sixteen years in travel:

  > Being in the tourism industry with two children on the autism spectrum, I felt it was important to become certified as a CATP. Hopefully, I can help [similar] families as they contemplate travel. Leaving home for an extended period can be extremely stressful—any little thing can have a massive impact on the family (either good or bad). Awareness and preparation are key elements to successful travel . . . [and] the training was extremely helpful [by offering] different viewpoints from the various guest

speakers, [though] most of the main points I was already aware
of due to my personal situation.

- **William Johnson is with Details by Bill LLC** of Germantown,
Wisconsin, and has been in travel for nineteen years. He pursued
his CATP designation when "a friend with a child on the spectrum
talked about difficulties traveling as a family. [The training] taught
us about autism and issues related to travel," with a key takeaway
being "the need to prepare kids for travel and [determine] what the
child will experience during the trip."

- **Sarah Marshall of TravelAble** in Midlothian, Virginia, has been
in travel for two years. During that time, she's helped at least fourteen neurodiverse families, some traveling multiple times. "As a
mother of a young man with autism, I felt the CATP designation
gave me an authority to act on the behalf of all of my clients. It has
helped me to view destinations in a different way, to really poke
and prod into how they can be more inclusive of all families. The
one thing that was staggering was the amount of potential money
floating around that many destinations could harness if they would
simply make a commitment to accessibility."

- **Cathi Marcheskie Maziarz of Glass Slipper Concierge**, in
Douglassville, Pennsylvania, has worked in travel for nine years. She
pursued her designation because "I know many people who have
children somewhere on the spectrum. I see what they go through in
everyday life. As a travel professional, when I became aware of the
CATP program, it was a natural fit for me. I wanted to help families
have a stress-free vacation. Taking classes to receive the certification
opened my eyes to situations and things that I was not aware of
before." She believes she has helped ten families with children on
the spectrum but "it could have been more. Some families do not
tell me, and I find out after they have traveled."

- **Olivia McKerley is with Elite Events & Tickets** in Grovetown,
Georgia. She's worked in travel for nearly seven years:

  I wanted to add the CATP certification to become a more well-
  rounded family travel professional. I have always loved working
  with autistic children when I taught elementary school. . . .
  My [own] children have food allergies which means we always

think and plan ahead to meet their dietary needs. However, families with autistic children have many more factors to consider when planning their vacations. They need enough information to create a Social Story for their child, make sure their vacation destination will be appropriate for their child's needs to prevent them from being overstimulated, and more. They can't just show up and wing it like other families. . . . I was able to make recommendations on autism-friendly vacation options to a former Walt Disney World client who was trying to decide where to vacation this year that wasn't a large theme park. She didn't want to fly, and cruises were not currently running, so I sent her a good list of options.

- **Suzanne Polak is with AAA** in Southington, Connecticut. During her sixteen years in travel, she estimates she's helped fifty families on the spectrum. She pursued the CATP designation "because I believe it is important to help parents pick out the right destination for their needs and ensure they and their children are going to have a great vacation." Key training takeaway? "That clients' needs are specific to them. No two clients' needs are the same."

- **Brenda Revere of C2C Escapes Travel** in Bedford, Indiana, has spent the past seven years in travel. She pursued her CATP designation because "I have a niece that is autistic and saw this as a market that is underserved. [I wanted to] understand the difficulties of traveling with autistic children [and have] learned which travel suppliers, hotels, cities, and theme parks help families ease their difficulties while traveling. [Now I] know the different types of autism and where to reach out for information."

- **Holly Roberts of Frosch Travel—The Travel Center** in Hanover, Massachusetts, spent twenty-nine years in travel and eighteen years in the airline industry:

  > I wanted to get my CATP because for twenty-four years, I have booked travel for special needs young adults attending Cardinal Cushing School, a school for special needs in Hanover, MA. Many families with a child on the spectrum believe their child cannot travel due to their disabilities. . . . They feel that arriving at the airport, going through TSA, getting on the plane, and then having to sit still for several hours is beyond what their

child can handle. [However,] 93 percent of parents feel that if autism [tourism] services were available, they would travel. So, parents need to have a CATP assist them with their travel needs. The key takeaway from the training is [the importance of] understanding all the aspects of a child on the spectrum. Having this knowledge allows me to understand what type of travel would be best for [that] child.

- **Angela Romans is with Rejuvia Travel** in South Bend, Indiana:

    As a newer travel advisor, I'd like to be able to help any client who needs travel planning assistance. I thought it would be a worthwhile endeavor to obtain the CATP designation so I could gain a better understanding of the condition, in addition to autism-friendly destinations, activities, and programs which are available to help make traveling easier. . . . This course offers a wealth of extremely useful information, [such as] the characteristics of autism spectrum disorder, how the senses can be affected for people on the spectrum, and how to communicate with someone who has a sensory disorder. There were also important subjects such as recognizing anxiety/social anxiety, how to deal with a meltdown, and tips, tricks, and programs available. . . . I now have . . . more confidence that I can [plan travel] for those with autism or other spectrum disorders, and a greater empathy for those living on the spectrum.

- **Sarah Salter is with Travelmation, dba Sarah Travels Happy** in Fayetteville, Georgia. Sarah has worked in travel for three years and pursued her CATP designation because

    as a special needs mom to three, I know how challenging it is to provide your children with the life-enhancing benefits of travel. I know what it's like to get the "side eye" when your kid uses a stroller as a wheelchair, or when an overzealous TSA agent swabs each can of medical formula causing you to miss your flight, or the terror of having your autistic child elope in a crowded theme park. I know the grief of thinking that some things are no longer possible for your child and your family, so you put those dreams away in a box on a shelf. If I can help other families along a journey that took our family almost fifteen years to complete, so that they [can] open that box of shuttered dreams, then why wouldn't I? I joke with folks that with me you get both street smarts (special needs mom) and book smarts (CATP designation).

Salter says she works closely with her local school district as well as several special needs charities at resource fairs to offer help to families hoping to travel.

- **Marian Speid of Marian Speid Dream Vacations** in Fair Lawn, New Jersey, has been involved with travel for six years. She studied for her CATP when

> I heard from friends about the difficulties of traveling with children who have various issues, [like being] stared at or having comments made because the children acted out. When I learned there were options for families with special needs, I wanted to learn more and discovered the certification. . . . Many people think it cannot be done or it is extremely hard to do. . . . There is also a concern that many families have, that they will become a distraction to other travelers, or they will be pointed at and made fun of. . . . Traveling with children is an adventure whether on the spectrum or not. It requires planning and preparation. Neurotypical children need to be entertained and kept busy. That may be easier when the kids can amuse themselves—a Game Boy, book, or some other toy. Of course, a child with ASD requires a little more planning depending on [where they fall] on the spectrum. They may have a *special* toy or book that amuses them. Keeping them occupied is essential.

- **Belinda Stull, owner/manager of Travel Leaders/All About Travel, LLC**, in Hagerstown, Maryland, has spent thirty-seven years working in travel. She chose to get her CATP designation because "my grandson was diagnosed with autism two years ago. I realized I could help him and others on the spectrum enjoy travel by pinpointing destinations that would meet everyone's needs. [The training taught me] that all children, neurotypical or not, are unique. While there may be challenges for children on the spectrum, each of these children have special interests, talents, and abilities that need to be considered to create the best travel experience for them."

- **Michelle St. Pierre is with Modern Travel Professionals** in Yardley, Pennsylvania. Michelle says that the CATP training reinforced practices that she already knew from working with families. "My business is focused on family travel, so all the seven years I've been doing this, I can think of five to seven families that I specifically helped with accommodations or deviations due to their child or children being on the spectrum, but there have been many more who benefited from my standard advice."

- **Sally J. Sutter of Mermaid Travel** in Canal Winchester, Ohio, has spent more than fifteen years in travel:

  > When my grandson (who is now almost twenty-one) was about one, he started exhibiting signs that he might have autism. [But] after extensive testing, it was determined that he had apraxia. In the meantime, my daughter-in-law's five-year-old grandson was diagnosed with autism. . . . I wanted to learn more about . . . how I could help [spectrum] families. I am special needs certified to help those with mobility, hearing, breathing, and sight disabilities, so it seemed like a natural extension. [The training has helped me] mainly to overcome my stereotypical thinking about children with autism: that they can control their behavior, or they are acting out on purpose, or that parents should have more control. Getting this designation educated me on the reality of those dealing with autism, [such as] the different spectrums, how many are affected, ways to help a child cope, the impact on family finances, and the job opportunities for those on the spectrum.

- **Nicole Thibault of Magical Storybook Travels** in Fairport, New York, is the parent of three sons—one on the autism spectrum, one with childhood apraxia of speech and sensory processing disorder, and one neurotypical. She is also a Certified Autism Travel Agent, an author, and an autism travel video producer. She's been traveling with her children since the youngest was six months old. Her main apprehensions? "Getting through the airport, flying, finding foods on his preferred meal list, public meltdowns, being unable to control the sensory environment." She says her main objectives were "relaxation, exposing the children to the world and conducting business, while having fun and making family memories together." Both her desire and her challenge were to "be just like every other 'typical' family on vacation," but she acknowledges that the preparation is different: "We've become much more organized when we travel. We have a timed schedule, know where we're staying, where we're eating, have [advance] reservations made at those restaurants, and have our daily activities planned. I think that benefits everyone in the family."

Her first inkling about her son's diagnosis occurred while her family was on the road:

> It was on a trip to Walt Disney World that we understood there was something different about our son. We had been to Walt Disney World before, and he always enjoyed the characters, the parades, and the attractions. On this particular trip, when he was 2.5 years old, everything upset him. The noises caused meltdowns, he didn't like the different smells of food, and waiting in lines made him extremely upset. After this trip, we knew we needed to find out the cause for all the meltdowns and sought a clinical evaluation. Weeks later, they confirmed what we already knew in our hearts, that he has autism.

- **Nicole Graham Wallace is with Trendy Travels by Nicole—Dream Vacations** in McDonough, Georgia:

> I decided to get the CATP designation because I have a son who has profound and severe hearing loss and while he is not on the spectrum, I thought it would be difficult to travel with him because of his deafness. I realized that it . . . is [actually] quite easy if you do your research and request any necessary assistance due to his hearing loss. I want other parents to understand they too can travel and not be limited. They are able to enjoy a memorable family and/or friends' vacation. [The training] has helped me to understand that there is not a one-size-fits-all solution for individuals on the spectrum. Individuals are different in their own ways, and we must understand those differences and how to best assist with planning a vacation that is stress free, enjoyable, and fun for everyone.

- **Cathy Winter of EnVision Travel** in Denver, North Carolina, chose to pursue her CATP designation because "I have a passion for autism and special needs and [wanted to learn] to handle this niche and assess the situation." At the time of writing, she had handled two spectrum families after thirty-five years of helping families with neurotypical children.

- **Starr Wlodarski is with Cruise Planners—Unique Family Adventures** in Perrysburg, Ohio. She's been in travel for eight years and pursued her CATP designation "to gain more knowledge above my own experience of traveling for twenty years with my son." She feels the training has opened doors to autism organizations, though she wishes it offered more information on suppliers other than Beaches Resorts, which she believes targets travelers with higher budgets.

- **Tara Woodbury of Escape into Travel** in Leland, North Carolina, is both a travel agent and the parent of a child on the spectrum. She's been in travel for three years and studied for her CATP designation because

  > I was always interested in helping families with special needs and medical issues [like ours] travel more easily. My daughter has complex medical needs and ADHD, so traveling with her isn't easy. I have several friends whose children are on the spectrum; I knew they had some of the same needs and concerns as I did when traveling but also some vastly different needs. So, I wanted to learn everything I could to advise them. I feel having this designation sets me apart . . . [because] I can offer knowledgeable advice. When families with someone on the spectrum learn that I have a child on the spectrum too, and that I also hold the Certified Autism Travel Advisor designation, they know I will take care of their family. . . . I never really realized how many people are on the spectrum [but] you wouldn't even know it. The folks I know personally are more on the severe end.

MENTAL HEALTH EXPERTS CONSULTED

- **Tony Attwood, Ph.D.**, is a British clinical psychologist and the acknowledged world expert on therapy with ASD clients. A professor at both Griffith University in Australia and Strathclyde University in Scotland, he's worked with an estimated eight thousand clients with some form of ASD diagnosis. He is the author of innumerable peer-reviewed studies and book chapters on autism, Asperger's syndrome, and ASD. He's also published fifteen books in the field with sales topping one million, and titles translated into twenty-eight languages. These include *CBT to Help Young People with Asperger's Syndrome to Understand and Express Affection: A Manual for Professionals* (2013) and *Exploring Depression and Beating the Blues: A CBT Self-Help Guide to Understanding and Coping with Depression in Asperger's Syndrome* (2016). He resides in Queensland, Australia.

- **Janeen Herskovitz** is a licensed mental health counselor in the state of Florida and owner of Puzzle Peace Counseling, LLC. Her son, Ben, was diagnosed with profound autism in 2001, and since then it has been her life's work to help other parents with similar struggles. She earned her bachelor's degree in special education from Rowan University in New Jersey (1995) and her master of arts in mental health counseling from Webster University (2010). She has been helping parents of children with autism spectrum issues since 2006. Additionally, she is trained in the Son-Rise therapy method for autism from the Option Institute and served as an ambassador parent for the program for over four years.

- **Ellen B. Littman, Ph.D.,** is a clinical psychologist licensed in New York State. Educated at Brown and Yale Universities, the clinical psychology doctoral program of Long Island University, and the Albert Einstein College of Medicine, she has been involved with the field of attentional disorders for over thirty years. She has been described by the American Psychological Association as "a pioneer in the identification of gender differences in ADHD." Dr. Littman is co-author of the book *Understanding Girls with ADHD* and is a contributing author of several books including *The Hidden Side of Adult ADHD*, *Understanding Women with ADHD*, and *Gender Differences in ADHD*.

NOTES

CHAPTER 1. EVERY FAMILY DESERVES TO TRAVEL

1. Juliet Perrachon. "Just When You Think Traveling with Kids Can't Get Any Worse, It Does." Scarymommy.com, last modified June 24, 2021. https://www.scary mommy.com/travel-with-kids-horror-stories/.

2. Ellen Edmonds. "AAA: Nearly 100 Million Americans Will Embark on Family Vacations this Year." AAA Newsroom.com, March 21, 2019. https://newsroom.aaa .com/2019/03/100-million-americans-will-embark-on-family-vacations/.

3. Dr. Lynn Minnaert. "U.S. Family Travel Survey 2017," by the Jonathan M. Tisch Center for Hospitality and Tourism, NYU School of Professional Studies, and the Family Travel Association, January 2018. http://2ks7od1dgl654aorxx98q3qu-wpengine .netdna-ssl.com/wp-content/uploads/2018/01/FTA.NYU_.2017-Consumer-Family -Travel-Survey.pdf.

4. Matt Turner. "Stats: 65 Percent of U.S. Travelers Looking Forward to Traveling Again." Travel Agent Central, June 3, 2020. https://www.travelagentcentral.com /your-business/stats-65-percent-u-s-travelers-looking-forward-to-traveling-again.

5. Edmonds. "100 Million Family Vacations."

6. Matthew J. Maenner et al. "Prevalence and Characteristics of Autism Spectrum Disorder among Children Aged 8 Years—Autism and Developmental Disabilities Monitoring, 11 Sites, United States, 2018." Centers for Disease Control and Prevention, December 3, 2021. https://www.cdc.gov/mmwr/volumes/70/ss/ss7011a1.htm.

7. U.S. Department of Health and Human Services/Health Resources and Services Administration/Maternal and Child Health Bureau. "The National Survey of Children with Special Health Care Needs." Last accessed October 20, 2021. https://mchb.hrsa .gov/chscn/pages/prevalence.htm.

8. Maggie Fox. "First US Study of Autism in Adults Estimates 2.2% Have Autism Spectrum Disorder." Last modified May 11, 2020. https://www.cnn.com/2020/05/11 /health/autism-adults-cdc-health/.

9. Pilar Clark. "Tips for Traveling with Kids on the Autism Spectrum." Travelmamas .com, April 7, 2016. https://travelmamas.com/traveling-with-kids-autism/.

CHAPTER 2. TRAVELING WITH
ANY CHILD CAN BE CHALLENGING

1. Alyson Long. "Travel Horror Stories." World Travel Family Travel Blog, last updated September 2, 2020. https://worldtravelfamily.com/travel-horror-stories/.

2. American Psychiatric Association. "What Is Autism Spectrum Disorder Fact Sheet." August 2021. https://www.psychiatry.org/patients-families/autism/what-is-autism-spectrum-disorder.

3. American Psychiatric Association. "What Is ASD?" https://www.psychiatry.org/patients-families/autism/what-is-autism-spectrum-disorder.

4. Centers for Disease Control and Prevention. "Attention-Deficit Hyperactivity Disorder (ADHD) Fact Sheet." Last reviewed September 23, 2021. https://www.cdc.gov/ncbddd/adhd/data.html.

5. Children and Adults with Attention-Deficit/Hyperactivity Disorder. "General Prevalence of ADHD." 2021. https://chadd.org/about-adhd/general-prevalence/.

6. Attention Deficit Disorder Association. "ADHD—The Facts." 2021. https://add.org/adhd-facts/.

7. Kimberly Holland. "Can Bipolar Disorder and Autism Co-Occur?" Healthline, last modified July 26, 2021. https://www.healthline.com/health/bipolar-and-autism.

8. Depression and Bipolar Support Alliance. "Bipolar Disorder: Diagnosis and Symptoms Fact Sheet." 2021. https://www.dbsalliance.org/education/bipolar-disorder/diagnosis/.

CHAPTER 3. THE GROWING FOCUS ON TRAVELING
WITH CHILDREN ON AND OFF THE SPECTRUM

1. Michele Bigley. "Sometimes Traveling with Kids Sucks." Huffpost, The Blog, last modified December 6, 2017. https://www.huffpost.com/entry/sometimes-traveling-with-_b_3824649.

2. National Institute of Mental Health. "Autism Spectrum Disorder Fact Sheet." 2021. https://www.nimh.nih.gov/health/topics/autism-spectrum-disorders-asd/index.shtml.

3. National Autism Association. "Autism Fact Sheet." https://nationalautismassociation.org/resources/autism-fact-sheet/.

4. Autism Speaks. "Sensory Issues." 2021. https://www.autismspeaks.org/sensory-issues.

5. Katherine G. Hobbs. "Severe Low-Functioning Autism, What Sets It Apart?" *Autism Parenting Magazine*, May 18, 2021. https://www.autismparentingmagazine.com/low-functioning-autism/.

6. G. A. Alvares, et al. "The Misnomer of 'High Functioning Autism': Intelligence Is an Imprecise Predictor of Functional Abilities at Diagnosis." *Autism*, January 2020, 221–32.

7. Attention Deficit Disorder Association. "ADHD—The Facts." https://add.org/adhd-facts/.

8. Dr. Elizabeth Harstad. "I've Heard that Autism and ADHD Are Related, Is That True?" Understood.org, 2021. https://www.understood.org/en/learning-thinking-differences/child-learning-disabilities/add-adhd/ive-heard-that-autism-and-adhd-are-related-is-that-true.

9. Harstad. "I've Heard that Autism and ADHD Are Related?"

10. Centers for Disease Control and Prevention. "Data and Statistics on Children's Mental Health." Last reviewed March 22, 2021. https://www.cdc.gov/childrensmentalhealth/data.html.

11. Children and Adults with Attention-Deficit/Hyperactivity Disorder. "General Prevalence ADHD." https://chadd.org/about-adhd/general-prevalence/.

12. Ann M. Reynolds and Beth A. Malow. "Sleep and Autism Spectrum Disorders." National Jewish.org, 2011. https://www.nationaljewish.org/NJH/media/pdf/Meltzer%20References/Reynolds-(2011)-Sleep-and-autism-spectrum-disorders.pdf.

13. Julie A. Fast. "Sleep Disturbance: The #1 Reason Travel Triggers Bipolar Disorder Mood Swings." BPHope.com, last modified May 16, 2019. https://www.bphope.com/blog/sleep-disturbance-travel-trigger-bipolar-mood-swings/.

14. Lonely Planet, "Family Travel Handbook" (2020), 6.

15. Travel Community (Private Group by Scott's Cheap Flights), Facebook. "What Are Some of the Biggest Problems You've Had When Traveling with Your Kids?" December 12, 2020. https://www.facebook.com/groups/scftravelcommunity/permalink/1313555022340779

16. Family Travel Association. "Vision and Mission." 2021. https://familytravel.org/mission/.

17. Taylor Covington. "Traveling with Autism: How to Handle Safety, Transitions, and Time in Transit." The Zebra.com, last modified August 10, 2021. https://www.thezebra.com/resources/driving/traveling-autism.

18. International Board of Credentialing and Continuing Education Standards. "Family Travel Association (FTA) and (IBCCES) have Teamed Up to Ensure Travel Agents from All Over the Globe have Access to Training and Certification in the Special Needs Travel Market." 2021. https://ibcces.org/fta/.

19. Karthikeyan Ardhanareeswaran and Fred Volkmar. "Introduction." *Yale Journal of Biology and Medicine*, March 4, 2015. https://www.ncbi.nlm.nih.gov/pmc/articles/PMC4345536/.

20. International Board of Credentialing and Continuing Education Standards. "Family Travel Association (FTA) and (IBCCES) have Teamed Up."

21. International Board of Credentialing and Continuing Education Standards. "Family Travel Association (FTA) and (IBCCES) have Teamed Up."

22. Note: Several attempts to secure a full list of CARD-certified venues went unanswered so I am unable to provide it here.

23. U.S. Department of Health and Human Services/Health Resources and Services Administration/Maternal and Child Health Bureau. "The National Survey of Children with Special Health Care Needs." Last accessed October 20, 2021. https://mchb.hrsa.gov/chscn/pages/prevalence.htm.

CHAPTER 4. MODIFICATIONS CAN
HELP EVERY FAMILY'S TRAVEL

1. Rebecca Schrag Hershberg. "Travel and Toddlers and Tantrums, Oh My." PsychologyToday.com, December 17, 2019. https://www.psychologytoday.com/us/blog/little-house-calls/201912/travel-and-toddlers-and-tantrums-oh-my.

2. Beth Arky. "Tips for Traveling with Challenging Children." Childmind.org, 2021. https://childmind.org/article/tips-for-traveling-with-challenging-children/.

3. Lisa Lightner. "33 Visual Picture Schedules for Home/Daily Routines." A Day in Our Shoes, March 26, 2020. https://adayinourshoes.com/free-printable-visual-schedules-for-home-and-daily-routines/.

4. Big Sky Therapeutic Services, PLLC. "The Five Best Types of Toys for Children with Autistic Spectrum Disorder." Last reviewed January 1, 2022. https://bigskytherapy.com/blog/the-best-toys-for-children-and-adults-with-autism.

CHAPTER 5. STARTING SMALL

1. Nicole Thibault. "Are Autism Parents 'Cautious Travelers'?" Autismtravel.com, 2021. https://autismtravel.com/2018/02/02/are-autism-parents-cautious-travelers/.

2. Nicole Thibault. "The Zoo Is a Perfect Outing for Children with Autism." Autismtravel.com, 2021. https://autismtravel.com/2018/06/04/parent-perspective-the-zoo-is-a-perfect-outing-for-kids-with-autism/.

3. International Board of Credentialing and Continuing Education Standards. "Why Autism Friendly Should Never Be Confused with Autism Certified." 2021. https://ibcces.org/blog/2019/04/24/autism-friendly-autism-certified/.

4. International Board of Credentialing and Continuing Education Standards. n.d. "Certified Autism Centers." Accessed October 15, 2020. https://certifiedautismcenter.com/.

5. International Board of Credentialing and Continuing Education Standards. n.d. "Becoming an Advanced Certified Autism Center." Accessed October 15, 2020. https://ibcces.org/advanced-certified-autism-center/.

6. Amazon.com. "Chris Oler Author Page." Accessed October 18, 2021. https://www.amazon.com/Chris-Oler/e/B00FSMO970/ref=ntt_dp_epwbk_0.

CHAPTER 6. PRE-TRIP PREPARATION

1. Taylor Covington. "Traveling with Autism: How to Handle Safety, Transitions, and Time in Transit." The Zebra.com, last modified August 10, 2021. https://www.thezebra.com/resources/driving/traveling-autism.

2. Amy Seeley. "Be Prepared: Carry an Autism Survival Kit." Autism Society of North Carolina, October 13, 2015. https://www.autismsociety-nc.org/be-prepared-carry-an-autism-survival-kit/.

3. Covington. "Traveling with Autism."

4. Nicole Thibault. "Episode 124: Traveling with Kids on the Autism Spectrum." The Vacation Mavens.com, 2021. https://vacationmavens.com/124-autism-travel/.

5. Covington. "Traveling with Autism."

6. Nicole Thibault. "Are Autism Parents 'Cautious Travelers'?" Autismtravel.com, 2021. https://autismtravel.com/2018/02/02/are-autism-parents-cautious-travelers/.

7. Natasha Daniels. "Eight Things Every Parent Should Do When Traveling with Anxious Children." AT Parenting Survival, 2021. https://www.anxioustoddlers.com /traveling-with-anxious-children.

8. Covington. "Traveling with Autism."

9. Beth Arky. "Tips for Traveling with Challenging Children." Childmind.org, 2021. https://childmind.org/article/tips-for-traveling-with-challenging-children/.

10. Autism Inspiration.com, "What Is a Social Story?" Last accessed October 18, 2021. https://www.autisminspiration.com/members/304.cfm.

11. Alethea Mshar. "Special Needs Parenting: What is a Social Story and How Can I Get One?" Shield Healthcare, September 6, 2018. http://www.shieldhealthcare .com/community/grow/2018/09/06/special-needs-parenting-what-is-a-social-story -and-how-can-i-get-one/.

CHAPTER 7. FOREIGN VERSUS DOMESTIC?

1. International Board of Credentialing and Continuing Education Standards. "Ramon's Village Resort in Belize Is Now a Certified Autism Center." July 28, 2020. https://ibcces.org/blog/2020/07/28/ramons-village-resort/.

2. Juve Travel Peru. "Define How You Want to Experience Peru." 2021. https:// juvetravelperu.com/.

CHAPTER 8. NAVIGATING AIRLINE TRAVEL

1. Plane Finder. "Plane Spotting and Autism." Plane Finder, January 4, 2019. https://planefinder.net/about/news/plane-spotting-and-autism/.

2. Judy Antell. "Traveling with Children on the Autism Spectrum." Family Travel Forum, last accessed October 18, 2021. https://myfamilytravels.com/content/52306 -traveling-children-autism-spectrum.

3. Operation Autism: A Resource Guide for Military Families. "Travel." 2021. https://operationautism.org/transitions/travel/#1543437067753-19a08b6d-ff2f.

4. Operation Autism. "Travel."

5. Operation Autism. "Travel."

6. Autism Speaks. "Tips for Holiday Travel!" November 13, 2017. https://www .autismspeaks.org/blog/tips-holiday-travel.

7. Transportation Security Administration. "TSA PreCheck®" 2021. https://www .tsa.gov/precheck or https://universalenroll.dhs.gov/workflows?servicecode=11115V& service=pre-enroll.

8. U.S. Customs and Border Protection. "Global Entry." 2021. https://www.cbp .gov/travel/trusted-traveler-programs/global-entry.

9. Operation Autism. "Travel."

10. Operation Autism. "Travel."

11. Taylor Covington. "Traveling with Autism: How to Handle Safety, Transitions, and Time in Transit." The Zebra.com, last modified August 10, 2021. https://www .thezebra.com/resources/driving/traveling-autism.

CHAPTER 9. ROAD TRIPS—
NEGOTIATING CAR AND BUS TRAVEL

1. Jackie Renzetti. "For This Year's Summer Vacations, 97 Percent Will Travel by Car." Bring Me the News, June 29, 2020. https://bringmethenews.com/minnesota -news/for-this-years-summer-vacations-97-percent-will-travel-by-car.

2. Taylor Covington. "Traveling with Autism: How to Handle Safety, Transitions, and Time in Transit." The Zebra.com, last modified August 10, 2021. https://www .thezebra.com/resources/driving/traveling-autism.

3. Car Talk. "How Can I Get My Car as Quiet as Possible on the Inside?" Dear Car Talk Blog, April 1, 2004. https://www.cartalk.com/content/my-youngest-son-has -been-diagnosed-autism-one.

4. Jenna Murrell. "When You Should Pass on the Rental Car Seat." Safe Ride 4 Kids, October 6, 2016. https://saferide4kids.com/blog/why-you-should-pass-on-the -rental-car-seat/.

5. Covington. "Safety, Transitions, Time in Transit."

6. Covington. "Safety, Transitions, Time in Transit."

7. Operation Autism: A Resource Guide for Military Families. "Travel." 2021. https://operationautism.org/transitions/travel/#1543437067753-19a08b6d-ff2f.

CHAPTER 10. CHUGGING ALONG—TRAIN TRAVEL

1. Amanda Bennett. "What Is It About Autism and Trains?!" Autism Speaks, September 6, 2018. https://www.autismspeaks.org/expert-opinion/what-it-about-autism -and-trains-0.

2. Autism Academy for Education and Development. "Better Understanding the Connection Between Autism and Trains." March 25, 2015. https://www.aaed.org /uncategorized/better-understanding-the-connection-between-autism-and-trains/.

3. Facebook. "Transport Sparks" private group. 2021. https://www.facebook.com /groups/transportsparks.

4. Amtrak. "Amtrak Baggage Allowance." 2021. https://media.amtrak.com/wp -content/uploads/2015/10/UPDATED-Baggage-fact-sheet-7_5_16.pdf.

CHAPTER 11. SMOOTH SAILING—CRUISE TRAVEL

1. Taylor Covington. "Traveling with Autism: How to Handle Safety, Transitions, and Time in Transit." The Zebra.com, last modified August 10, 2021. https://www.thezebra.com/resources/driving/traveling-autism.

2. Ida Keiper and Jesemine Jones. "A Travel Resource for Parents of Children with Special Needs." Family Travel Association, 2021. https://familytravel.org/travel-resource-parents-children-with-special-needs/.

3. Keiper and Jones. "A Travel Resource for Parents of Children with Special Needs."

4. Royal Caribbean International. "Autism Friendly Ships." 2021. https://www.royalcaribbean.com/experience/accessible-cruising/autism-friendly-ships.

5. Autism on the Seas. 2021. https://autismontheseas.com/.

6. Autism on the Seas. Testimonial page. 2021. Facebook. https://www.facebook.com/AutismontheSeas.Testimonials.

7. Autism on the Seas. 2021. https://autismontheseas.com/.

8. Six Suitcase Travel. "How to Rent a Houseboat for a Fantastic Big Family Vacation." 2021. https://sixsuitcasetravel.com/blog/how-to-rent-a-houseboat-for-a-fantastic-big-family-vacation/.

9. Cotton Cove Resort Marina. "Tips for Renting a Houseboat with Kids." July 22, 2020. https://cottonwoodcoveresort.com/tips-for-renting-a-houseboat-with-kids/.

10. Kiana Tom. "17 Tips You Must Know Before Renting a Houseboat." 2021. https://kiana.com/17-tips-must-know-before-renting-a-houseboat/.

11. Houseboating.org. "Top 5 Houseboating Safety Tips." 2021. https://www.houseboating.org/Top-5-Houseboating-Safety-Tips.

12. Six Suitcase Travel. "How to Rent a Houseboat."

CHAPTER 12. STAYING THE COURSE—
VARIOUS ACCOMMODATIONS OPTIONS

1. Nicole Thibault. "Episode 124: Traveling with Kids on the Autism Spectrum." The Vacation Mavens.com. 2021. https://vacationmavens.com/124-autism-travel/.

2. Taylor Covington. "Traveling with Autism: How to Handle Safety, Transitions, and Time in Transit." The Zebra.com, last modified August 10, 2021. https://www.thezebra.com/resources/driving/traveling-autism.

3. Covington. "Safety, Transitions, Time in Transit."

4. Covington. "Safety, Transitions, Time in Transit."

5. Daniels. "Eight Things When Traveling with Anxious Children."

6. Airbnb. "Don's Bay Getaway." 2021. https://www.airbnb.com/rooms/44195690?source_impression_id=p3_1595499742_7uOZarNATGCX5FtR.

7. International Board of Credentialing and Continuing Education Standards. "Beaches Becomes the First Resort Company in the World to Attain Advanced Certi-

fied Autism Center Credential." April 30, 2019. https://ibcces.org/blog/2019/04/30/beaches-resorts-becomes-first-resort-company-in-the-world-to-attain-advanced-certified-autism-center-credential/.

8. Beaches in the News. "Beaches Resorts Elevates Its Commitment to Inclusivity in Celebration of Autism Acceptance Month." Beaches.com, April 13, 2021. https://news.beaches.com/article/1572/.

9. Trey Duling. "Interview about Autistic Friendly Vacation Homes." Orlando Vacations.com, March 20, 2021. https://www.orlandovacation.com/autistic-friendly-vh-interview/.

10. Gene Kilgore Top 50 Ranches. "Therapeutic Horseback Riding for Kids with Autism." July 14, 2015. https://www.top50ranches.com/blog/therapeutic-horseback-riding-for-kids-with-autism.

CHAPTER 13. THE GREAT OUTDOORS— CAMPING, RVs, AND MORE

1. Angela Zizak. "An Autism Guide to Camping." Your Autism Guide, May 16, 2019. https://www.yourautismguide.com/2019/05/16/an-autism-guide-to-camping/.

2. Wigwam Holidays. "Glamping and Camping Tips for Families with Autistic Kids." 2018. https://blog.wigwamholidays.com/glamping-camping-tips-for-families-with-autistic-kids.

3. Ramah Darom. "Camp Yofi." 2021. https://www.ramahdarom.org/ramah-darom-jewish-summer-camp/summer-offerings/camp-yofi/.

4. Susan Kabot. "Camp Yofi: A Family Camp for Children with Autism." Reprinted by the American Camp Association. July 2009. https://www.acacamps.org/resource-library/camping-magazine/camp-yofi-family-camp-children-autism.

5. Dennis Thompson. "Autism Greatly Boosts Risk of Drowning." Health-Day News/WebMD, March 21, 2017. https://www.webmd.com/brain/autism/news/20170321/autism-greatly-boosts-kids-injury-risk-especially-for-drowning.

6. Tabatha Philen. "Camping with Special Needs Kids." Meet Penny, July 16, 2016. https://www.meetpenny.com/2016/07/camping-special-needs-children-autism/.

7. Zizak. "An Autism Guide to Camping."

8. Angela Zizak. "National Park Access Pass for Autism." Your Autism Guide, May 10, 2019. https://www.yourautismguide.com/2019/05/10/national-park-access-pass-for-autism/.

9. J. R. Reed. "How Geocaching Works for the Autistic Me." The Good Men Project, February 23, 2019. https://goodmenproject.com/featured-content/geocaching-works-for-autistic-me-jrrd/.

10. Erin Clemens. "How a GPS Game Helps Me as Someone on the Autism Spectrum." The Mighty, September 22, 2017. https://themighty.com/2017/09/gps-game-geocaching-autism-spectrum.

11. Matt Buxton. "Nine-Year-Old with Autism Finds Connection to Family While Geocaching." The Oregonian, last updated January 10, 2019. https://www.oregonlive.com/gresham/2010/12/autistic_9-year-old_boy_finds.html.

CHAPTER 14. TO SEE OR NOT TO SEE—
TOURS AND VENUES

1. Hend M. Hamed. "Tourism and Autism: An Initiative Study for How Travel Companies Can Plan Tourism Trips for Autistic People." *American Journal of Tourism Management*. 2013;2(1):1–14.

2. Taylor Covington. "Traveling with Autism: How to Handle Safety, Transitions, and Time in Transit." The Zebra.com, last modified August 10, 2021. https://www .thezebra.com/resources/driving/traveling-autism.

3. Alex Manson-Smith. "Why Children with Autism Love Trains (and the 10 Best U.K. Rail Adventures for Families)." *The Telegraph*, June 19, 2018. https://www.tele graph.co.uk/travel/family-holidays/best-train-attractions-in-uk-and-why-kids-with -autism-like-trains/.

4. Roxanna Swift. "Roller Coaster Conqueror Logan Joiner, on the Autism Spectrum, Helps Others Overcome Their Fears." WCPO-9, Cincinnati, April 24, 2017. https://www.wcpo.com/news/insider/logan-joiner-addresses-his-fears-and-those-of -others-on-the-autism-spectrum-by-riding-and-reviewing-roller-coasters.

5. Nicole Thibault. "Episode 124: Traveling with Kids on the Autism Spectrum." The Vacation Mavens.com. 2021. https://vacationmavens.com/124-autism-travel/.

6. Mark Hutten. "Kids on the Autism Spectrum and Amusement Parks: Avoiding Overstimulation." My ASD Child, May 2019. https://www.myaspergerschild .com/2019/05/kids-on-autism-spectrum-and-amusement.html.

7. Hutten. "Kids on the Autism Spectrum and Amusement Parks."

8. Applied Behavioral Analysis Programs Guide. "30 Best Autism-Friendly Vacation Ideas." June 2019. https://www.appliedbehavioranalysisprograms.com/30-best -autism-friendly-vacation-ideas/.

9. International Board of Credentialing and Continuing Education Standards. "What is an IBCCES Accessibility Card (IAC)?" 2021. https://ibcces.org/ac/.

10. Wilderness Inquiry. "Our Mission: Everyone Belongs in the Outdoors." 2021. https://www.wildernessinquiry.org/mission/.

11. Wilderness Inquiry. "Families Together." 2021. https://www.wildernessinquiry .org/programs/accessible-family-vacations/.

CHAPTER 15. TAKING THE BITE OUT
OF RESTAURANT DINING

1. Riv's Toms River Hub. "Welcome to Chase's Friend Zone." 2021. https:// rivstrhub.com/chases-friend-zone/

2. Mark Hutten. "Dining Out with Children on the Autism Spectrum: 20 Tips for Parents." My ASD Child, May 2013. https://www.myaspergerschild.com/2013/05 /dining-out-with-children-on-autism.html.

3. Taylor Covington. "Traveling with Autism: How to Handle Safety, Transitions, and Time in Transit." The Zebra.com, last modified August 10, 2021. https://www .thezebra.com/resources/driving/traveling-autism.

4. Hutten. "Dining Out with Children on the Autism Spectrum."

CHAPTER 16. OTHER SUGGESTED MODIFICATIONS

1. Janet Cole, et al. "How to Decide Between Autism Service Dogs or Therapy Dogs." Autism Speaks, July 15, 2016. https://www.autismspeaks.org/expert-opinion/service-dog-or-therapy-dog-which-best-child-autism.

2. Hannah Sampson. "8 Questions About Flying with Emotional Support Animals, Answered." *The Washington Post*, January 14, 2021. https://www.washingtonpost.com/travel/2021/01/14/emotional-support-animal-airlines-restrictions/.

3. All Things Cruise. "Cruise Line Pet Policies." 2021. https://allthingscruise.com/cruise-line-pet-policies/.

4. ESA Doctors. "Which Airlines Are Still Allowing Emotional Support Animals?" 2021. https://esadoctors.com/airlines-allowing-emotional-support-animals/.

5. Autism Help.org. "Autism Spectrum Disorders Fact Sheets/Tourette Syndrome." 2021. http://www.autism-help.org/comorbid-tourette-syndrome-autism.htm.

6. Tourette Association of America. "What Is Tourette." 2021. https://tourette.org/about-tourette/overview/what-is-tourette/.

7. Tourette Association of America. "Print an 'I Have TS' Card.'" 2021. https://tourette.org/about-tourette/overview/living-tourette-syndrome/print-ts-card/.

8. Marc Elliot. "'AA' Angles in the Sky: Flying with Tourette Syndrome." Friendship Circle, August 20, 2010. https://www.friendshipcircle.org/blog/2010/08/20/%E2%80%9Caa%E2%80%9Dngles-in-the-sky-flying-with-tourette-syndrome/.

9. Margarita Tartakovsky. "12 Travel Tips for People with Bipolar Disorder." PsychCentral, August 3, 2013. https://psychcentral.com/blog/12-travel-tips-for-people-with-bipolar-disorder.

10. RTOR.org. "Borderline Personality Disorder." 2021. https://www.rtor.org/borderline-personality-disorder.

11. Marlene Valle. "Traveling with Mental Illness: What Is It Really Like?" DeafinitelyWanderlust.com. May 2018. https://deafinitelywanderlust.com/2018/05/traveling-with-mental-health-illness-what-is-it-really-like/.

12. Taylor Covington. "Traveling with Autism: How to Handle Safety, Transitions, and Time in Transit." The Zebra.com, last modified August 10, 2021. https://www.thezebra.com/resources/driving/traveling-autism.

13. Covington. "Safety, Transitions, Time in Transit."

14. Nicole Thibault. "Episode 124: Traveling with Kids on the Autism Spectrum." The Vacation Mavens.com. 2021. https://vacationmavens.com/124-autism-travel/.

CHAPTER 17. SUGGESTED ITINERARIES

1. Visit Mesa. "Mesa, Arizona is the First-Ever Autism Certified City in the United States." 2021. https://www.visitmesa.com/mesaautismcertifiedtravel/.

2. Champion Autism Network. "CAN Card." 2021. https://championautismnetwork.com/can-card/.

3. Surfside Beach, South Carolina. "Autism Friendly." 2021. https://www.surfsidebeach.org/autism-friendly.

4. Lorena Armstrong-Duarte. "Minnesota Attractions that Cater to Kids with Autism." Explore Minnesota, 2021. https://www.exploreminnesota.com/article/min nesota-attractions-that-cater-to-kids-autism.

5. Blake Schuster. "PGA Tour's Billy Mayfair Says He Has Autism Spectrum Disorder." Bleacher Report, April 21, 2021. https://bleacherreport.com/articles/2940757 -pga-tours-billy-mayfair-says-he-has-autism-spectrum-disorder.

6. Kathleen O'Brien. "Kids with Autism Take a Swing at Golf, Courtesy of PGA Pro Ernie Els." NJ.com, last updated March 29, 2019. https://www.nj.com/health fit/2015/06/golf_and_autism.html.

7. Autism Travel. "Valley Oaks Golf Course, The First of Its Kind to Become Autism Certified." 2021. https://autismtravel.com/2021/12/02/valley-oaks-golf-course/.

8. *Golf Digest* Editors. "Best Golf Resorts for Families: Golf's Top 100 Resorts 2019." September 26, 2019. https://golf.com/travel/best-golf-resorts-families-top-100 -resorts/.

9. Kristine Dworkin. "Introducing Kids on the Autism Spectrum to Skiing." Trekaroo.com, March 22, 2018. https://blog.trekaroo.com/introducing-kids-on-the -autism-spectrum-to-skiing/.

10. Jodi Saeland (Mountain Mama). "Skiing/Boarding Best Therapy for Those with Disabilities." Ski Utah, March 22, 2017. https://www.skiutah.com/blog/authors/jodi /for-those-with-disabilities-skiing/.

11. Karin/Blogger. "Ski Resorts with Adaptive Skiing Programs." Special Needs Travel Mom, October 26, 2018. http://specialneedstravelmom.com/blog/ski-resorts -adaptive-skiing-programs/.

12. International Board of Credentialing and Continuing Education Standards. "Top 8 Travel Options That Include Water for Families with Autism." May 2, 2019. https:// ibcces.org/blog/2019/05/02/travel-options-water-families-autism/.

13. Megan Denny. "How Scuba Diving Can Help People with Autism." PADI, 2021. https://blog.padi.com/scuba-diving-can-help-people-autism/.

14. Alice Kay. "Scuba Diving: A Soothing Adventure." *Autism Parenting Magazine*, May 29. 2020. https://www.autismparentingmagazine.com/scuba-diving-a-soothing -adventure/.

15. Adwpadmin. "'Scuba for Autism' Program in Loves Park." WREX.com, April 6, 2018. https://wrex.com/2018/04/06/scuba-for-autism-program-in-loves-park/.

16. California Diver. "Benefits of Scuba Diving Shows Promise to Those with Autism." Calfiorniadiver.com. April 11, 2016. https://californiadiver.com/benefits-of -scuba-diving-shows-promise-to-those-with-autism-0410/.

17. Diveheart. "About Diveheart." 2021. https://www.diveheart.org/about-dive heart/.

18. Emily Laber-Warren. "The Benefits of Special Interests in Autism." Spectrum News, May 12, 2021. https://www.spectrumnews.org/features/deep-dive/the-bene fits-of-special-interests-in-autism/.

19. BAA Training. "Top 10 Aviation, Air, and Space Museums in the World." January 3, 2016. https://www.baatraining.com/top-10-aviation-air-and-space-museums -in-the-world/.

BIBLIOGRAPHY

Adwpadmin. "'Scuba for Autism' Program in Loves Park." WREX.com, April 6, 2018. https://wrex.com/2018/04/06/scuba-for-autism-program-in-loves-park/.

Airbnb. "Don's Bay Getaway." 2021. https://www.airbnb.com/rooms/44195690?source _impression_id=p3_1595499742_7uOZarNATGCX5FtR.

All Things Cruise. "Cruise Line Pet Policies." 2021. https://allthingscruise.com/cruise -line-pet-policies/.

Alvares, G.A., Bebbington, K., Cleary, D., Evans, K., Glasson, E.J., Maybery, M.T., Pillar, S., Uljarević, M., Varcin, K., Wray, J., and Whitehouse, A.J. "The Misnomer of 'High Functioning Autism': Intelligence Is an Imprecise Predictor of Functional Abilities at Diagnosis." *Autism*. January 2020, 221–32.

Amazon.com. "Chris Oler Author Page." Accessed October 18, 2021. https://www .amazon.com/Chris-Oler/e/B00FSMO970/ref=ntt_dp_epwbk_0.

American Psychiatric Association. "What Is Autism Spectrum Disorder Fact Sheet." August 2021. https://www.psychiatry.org/patients-families/autism/what-is-autism -spectrum-disorder.

Amtrak. "Accessible Travel Services." 2021. https://www.amtrak.com/accessible -travel-services.

Amtrak. "Amtrak Baggage Allowance." 2021. https://media.amtrak.com/wp-content /uploads/2015/10/UPDATED-Baggage-fact-sheet-7_5_16.pdf.

Antell, Judy. "Traveling with Children on the Autism Spectrum." Family Travel Forum, last accessed October 18, 2021. https://myfamilytravels.com/content/52306 -traveling-children-autism-spectrum.

Applied Behavioral Analysis Programs Guide. "30 Best Autism-Friendly Vacation Ideas." June 2019. https://www.appliedbehavioranalysisprograms.com/30-best -autism-friendly-vacation-ideas.

Apstein, Stephanie. "Billy Mayfair Reveals Autism Spectrum Disorder Diagnosis." *Sports Illustrated*, April 21, 2021. https://www.si.com/golf-archives/2021/04/21 /billy-mayfair-autism-spectrum-disorder-diagnosis-pga-champions-tour.

Ardhanareeswaran, Karthikeyan, and Volkmar, Fred. "Introduction." *Yale Journal of Biology and Medicine*, March 4, 2015. https://www.ncbi.nlm.nih.gov/pmc/articles /PMC4345536/.

Arky, Beth. "Tips for Traveling with Challenging Children." Childmind.org, 2021. https://childmind.org/article/tips-for-traveling-with-challenging-children/.

Armstrong-Duarte, Lorena. "Minnesota Attractions that Cater to Kids with Autism." Explore Minnesota, 2021. https://www.exploreminnesota.com/article/minnesota-attractions-that-cater-to-kids-autism.

Attention Deficit Disorder Association. "ADHD—The Facts." 2021. https://add.org/adhd-facts/.

Autism Academy for Education and Development. "Better Understanding the Connection Between Autism and Trains." March 25, 2015. https://www.aaed.org/uncategorized/better-understanding-the-connection-between-autism-and-trains/.

Autism Help.org. "Autism Spectrum Disorders Fact Sheets/Tourette Syndrome." 2021. http://www.autism-help.org/comorbid-tourette-syndrome-autism.htm.

Autism Inspiration.com. "What Is a Social Story?" Last accessed October 18, 2021. https://www.autisminspiration.com/members/304.cfm.

Autism on the Seas. 2021. https://autismontheseas.com/.

Autism on the Seas Testimonial Page. Facebook, 2021. https://www.facebook.com/AutismontheSeas.Testimonials.

Autism Speaks. "Assistance Dog Information." 2021. https://www.autismspeaks.org/assistance-dog-information.

Autism Speaks. "Brooklyn Junior Autistic Golfers Academy." Autism Speaks. July 16, 2016. https://www.autismspeaks.org/provider/brooklyn-junior-autistic-golfers-academy.

Autism Speaks. "Sensory Issues." 2021. https://www.autismspeaks.org/sensory-issues.

Autism Speaks. "Tips for Holiday Travel!" November 13, 2017. https://www.autismspeaks.org/blog/tips-holiday-travel.

Autism Travel. "Valley Oaks Golf Course, The First of Its Kind to Become Autism Certified." 2021. https://autismtravel.com/2021/12/02/valley-oaks-golf-course/.

BAA Training. "Top 10 Aviation, Air, and Space Museums in the World." January 3, 2016. https://www.baatraining.com/top-10-aviation-air-and-space-museums-in-the-world/.

Beaches in the News. "Beaches Resorts Elevates Its Commitment to Inclusivity in Celebration of Autism Acceptance Month." Beaches.com, April 13, 2021. https://news.beaches.com/article/1572/

Bennett, Amanda. "What Is It About Autism and Trains?!" Autism Speaks, September 6, 2018. https://www.autismspeaks.org/expert-opinion/what-it-about-autism-and-trains-0.

Big Sky Therapeutic Services, PLLC. "The Five Best Types of Toys for Children with Autistic Spectrum Disorder." Last reviewed January 1, 2022. https://bigskytherapy.com/blog/the-best-toys-for-children-and-adults-with-autism.

Bigley, Michele. "Sometimes Traveling with Kids Sucks." Huffpost, The Blog, last modified December 6, 2017. https://www.huffpost.com/entry/sometimes-traveling-with-_b_3824649.

Brooklyn Junior Autistic Golfers Association. "Voices from the Outside." Vimeo, last accessed 2021. https://vimeo.com/212905669.

Buxton, Matt. "Nine-Year-Old with Autism Finds Connection to Family While Geo-caching." *The Oregonian*, last updated January 10, 2019. https://www.oregonlive.com/gresham/2010/12/autistic_9-year-old_boy_finds.html.

California Diver. "Benefits of Scuba Diving Shows Promise to Those with Autism." Californiadiver.com, April 11, 2016. https://californiadiver.com/benefits-of-scuba-diving-shows-promise-to-those-with-autism-0410/.

Car Talk. "How Can I Get My Car as Quiet as Possible on the Inside?" Dear Car Talk Blog, April 1, 2004. https://www.cartalk.com/content/my-youngest-son-has-been-diagnosed-autism-one.

Centers for Disease Control and Prevention. "Attention-Deficit Hyperactivity Disorder (ADHD) Fact Sheet." Last reviewed September 23, 2021. https://www.cdc.gov/ncbddd/adhd/data.html.

Centers for Disease Control and Prevention. "Data and Statistics on Children's Mental Health." Last reviewed March 22, 2021. https://www.cdc.gov/childrensmentalhealth/data.html.

Champion Autism Network. "CAN Card." 2021. https://championautismnetwork.com/can-card/.

Children and Adults with Attention-Deficit/Hyperactivity Disorder. "General Prevalence of ADHD." 2021. https://chadd.org/about-adhd/general-prevalence/.

Clark, Pilar. "Tips for Traveling with Kids on the Autism Spectrum." Travelmamas.com, April 7, 2016. https://travelmamas.com/traveling-with-kids-autism/.

Clemens, Erin. "How a GPS Game Helps Me as Someone on the Autism Spectrum." The Mighty, September 22, 2017. https://themighty.com/2017/09/gps-game-geocaching-autism-spectrum.

Cole, Janet, Miller, Betty, and Gott, Rolanda Maxim. "How to Decide Between Autism Service Dogs or Therapy Dogs." Autism Speaks, July 15, 2016. https://www.autismspeaks.org/expert-opinion/service-dog-or-therapy-dog-which-best-child-autism

Cotton Cove Resort Marina. "Tips for Renting a Houseboat with Kids." July 22, 2020. https://cottonwoodcoveresort.com/tips-for-renting-a-houseboat-with-kids/.

Covington, Taylor. "Traveling with Autism: How to Handle Safety, Transitions, and Time in Transit." The Zebra.com, last modified August 10, 2021. https://www.thezebra.com/resources/driving/traveling-autism.

Daniels, Natasha. "Eight Things Every Parent Should Do When Traveling with Anxious Children." AT Parenting Survival, 2021. https://www.anxioustoddlers.com/traveling-with-anxious-children.

Denny, Megan. "How Scuba Diving Can Help People with Autism." PADI, 2021. https://blog.padi.com/scuba-diving-can-help-people-autism/.

Depression and Bipolar Support Alliance. "Bipolar Disorder: Diagnosis and Symptoms Fact Sheet." 2021. https://www.dbsalliance.org/education/bipolar-disorder/diagnosis/.

Diveheart. "About Diveheart." 2021. https://www.diveheart.org/about-diveheart/.

Diveheart. "Instructor Directory." 2021. https://diveheart.org/instructor-directory.

Duling, Trey. "Interview about Autistic Friendly Vacation Homes." OrlandoVacations .com, March 20, 2021. https://www.orlandovacation.com/autistic-friendly-vh -interview/.

Dworkin, Kristine. "Introducing Kids on the Autism Spectrum to Skiing." Trekaroo .com, March 22, 2018. https://blog.trekaroo.com/introducing-kids-on-the-autism -spectrum-to-skiing/.

Edmonds, Ellen. "AAA: Nearly 100 Million Americans Will Embark on Family Vacations this Year." AAA Newsroom.com, March 21. 2019. https://newsroom.aaa .com/2019/03/100-million-americans-will-embark-on-family-vacations/.

Elliot, Marc. "'AA'ngles in the Sky: Flying with Tourette Syndrome." Friendship Circle, August 20, 2010. https://www.friendshipcircle.org/blog/2010/08/20/%E2% 80%9Caa%E2%80%9Dngles-in-the-sky-flying-with-tourette-syndrome/.

ESA Doctors. "Which Airlines Are Still Allowing Emotional Support Animals." 2021. https://esadoctors.com/airlines-allowing-emotional-support-animals/.

Facebook. "Transport Sparks" private group. 2021. https://www.facebook.com/groups /transportsparks.

Family Travel Association. "Vision and Mission." 2021. https://familytravel.org/mission/.

Fast, Julie A. "Sleep Disturbance: The #1 Reason Travel Triggers Bipolar Disorder Mood Swings." BPHope.com, last modified May 16, 2019. https://www.bphope .com/blog/sleep-disturbance-travel-trigger-bipolar-mood-swings/.

Finch, Pete. "Our Top 10 Destinations for Family Golf Trips." *Golf Digest*, December 22, 2015. https://www.golfdigest.com/story/our-top-10-destinations-for-family -golf-trips.

Fox, Maggie. "First US Study of Autism in Adults Estimates 2.2% Have Autism Spectrum Disorder." Last modified May 11, 2020. https://www.cnn.com/2020/05/11 /health/autism-adults-cdc-health/.

Garey, Juliann. "DBT: What Is Dialectical Behavior Therapy." Child Mind Institute, 2021. https://childmind.org/article/dbt-dialectical-behavior-therapy/.

Gene Kilgore Top 50 Ranches. "Therapeutic Horseback Riding for Kids with Autism." July 14, 2015. https://www.top50ranches.com/blog/therapeutic-horseback-riding -for-kids-with-autism.

Golf Digest Editors. "Best Golf Resorts for Families: Golf's Top 100 Resorts 2019." September 26, 2019. https://golf.com/travel/best-golf-resorts-families-top -100-resorts/.

Harstad, Elizabeth. "I've Heard that Autism and ADHD Are Related. Is That True?" Understood.org, 2021. https://www.understood.org/en/learning-thinking-differ ences/child-learning-disabilities/add-adhd/ive-heard-that-autism-and-adhd-are-re lated-is-that-true.

Hamed, Hend M. "Tourism and Autism: An Initiative Study for How Travel Companies Can Plan Tourism Trips for Autistic People." *American Journal of Tourism Management.* 2013;2(1):1–14.

Hershberg, Rebecca Schrag. "Travel and Toddlers and Tantrums, Oh My." Psychology Today.com, December 17, 2019. https://www.psychologytoday.com/us/blog/little -house-calls/201912/travel-and-toddlers-and-tantrums-oh-my.

Hobbs, Katherine G. "Severe Low-Functioning Autism, What Sets It Apart?" *Autism Parenting Magazine*, May 18, 2021. https://www.autismparentingmagazine.com/low-functioning-autism/.

Holland, Kimberly. "Can Bipolar Disorder and Autism Co-Occur?" Healthline, last modified July 26, 2021. https://www.healthline.com/health/bipolar-and-autism.

Houseboating.org. "Top 5 Houseboating Safety Tips." 2021. https://www.houseboating.org/Top-5-Houseboating-Safety-Tips.

Hutten, Mark. "Dining Out with Children on the Autism Spectrum: 20 Tips for Parents." My ASD Child, May 2013. https://www.myaspergerschild.com/2013/05/dining-out-with-children-on-autism.html.

Hutten, Mark. "Kids on the Autism Spectrum and Amusement Parks: Avoiding Overstimulation." My ASD Child, May 2019. https://www.myaspergerschild.com/2019/05/kids-on-autism-spectrum-and-amusement.html.

International Board of Credentialing and Continuing Education Standards. "Beaches Becomes the First Resort Company in the World to Attain Advanced Certified Autism Center Credential." April 30, 2019. https://ibcces.org/blog/2019/04/30/beaches-resorts-becomes-first-resort-company-in-the-world-to-attain-advanced-certified-autism-center-credential/.

International Board of Credentialing and Continuing Education Standards. "Becoming an Advanced Certified Autism Center." Accessed October 15, 2020. https://ibcces.org/advanced-certified-autism-center/.

International Board of Credentialing and Continuing Education Standards. "Family Travel Association (FTA) and IBCCES Have Teamed up to Ensure Travel Agents from All Over the Globe Have Access to Training and Certification in the Special Needs Travel Market." 2021. https://ibcces.org/fta/.

International Board of Credentialing and Continuing Education Standards. "Ramon's Village Resort in Belize Is Now a Certified Autism Center." July 28, 2020. https://ibcces.org/blog/2020/07/28/ramons-village-resort/.

International Board of Credentialing and Continuing Education Standards. "Top 8 Travel Options That Include Water for Families with Autism." May 2, 2019. https://ibcces.org/blog/2019/05/02/travel-options-water-families-autism/.

International Board of Credentialing and Continuing Education Standards. "What Is an IBCCES Accessibility Card (IAC)?" 2021. https://ibcces.org/ac/.

International Board of Credentialing and Continuing Education Standards. "Why Autism Friendly Should Never Be Confused with Autism Certified." 2021. https://ibcces.org/blog/2019/04/24/autism-friendly-autism-certified/.

International Board of Credentialing and Continuing Education Standards. "Certified Autism Centers." Accessed October 15, 2020. https://certifiedautismcenter.com/.

Isaacson, Rupert. *The Horse Boy: A Father's Miraculous Journey to Heal His Son.* New York: Little, Brown and Company, 2009.

Isaacson, Rupert. *The Long Ride Home: The Extraordinary Journey of Healing that Changes a Boy's Life (The Horse Boy Book 2).* Elgin, TX: Horse Boy Press, 2016.

Juve Travel Peru. "Define How You Want to Experience Peru." 2021. https://juvetravelperu.com/.

Kabot, Susan. "Camp Yofi: A Family Camp for Children with Autism." Reprinted by the American Camp Association, July 2009. https://www.acacamps.org/resource -library/camping-magazine/camp-yofi-family-camp-children-autism.

Karin/Blogger. "Ski Resorts with Adaptive Skiing Programs." Special Needs Travel Mom, October 26, 2018. http://specialneedstravelmom.com/blog/ski-resorts-adap tive-skiing-programs/.

Kay, Alice. "Scuba Diving: A Soothing Adventure." *Autism Parenting Magazine*, May 29. 2020. https://www.autismparentingmagazine.com/scuba-diving-a-soothing -adventure/.

Keiper, Ida, and Jones, Jesemine. "A Travel Resource for Parents of Children with Special Needs." Family Travel Association, 2021. https://familytravel.org/travel -resource-parents-children-with-special-needs/.

Kelleher, Suzanne Rowan. "America's Best Family-Friendly Golf Resorts." Trip Savvy, July 4, 2019. https://www.tripsavvy.com/americas-best-family-friendly-golf -resorts-3266633.

Koenig, Ronnie. "Travel and Autism: How My Family Found a Place That's Truly 'All-Inclusive.'" Better by Today, October 11, 2019. https://www.nbcnews.com/better /lifestyle/traveling-autistic-child-how-my-family-found-place-s-truly-ncna1063761.

Laber-Warren, Emily. "The Benefits of Special Interests in Autism." Spectrum News, May 12, 2021. https://www.spectrumnews.org/features/deep-dive/the-benefits-of -special-interests-in-autism/.

Learning Difficulties Resources. "Eat Out—Even with an Autistic Child." July 2011. http://learning-difficulties.blogspot.com/2011/07/eat-out-even-with-autistic-child .html.

Lightner, Lisa. "33 Visual Picture Schedules for Home/Daily Routines." A Day in Our Shoes, March 26, 2020. https://adayinourshoes.com/free-printable-visual-schedules -for-home-and-daily-routines/

Lonely Planet. *The Family Travel Handbook*. California: Lonely Planet Global Limited, 2020.

Long, Alyson. "Travel Horror Stories." World Travel Family Travel Blog, last updated September 2, 2020. https://worldtravelfamily.com/travel-horror-stories/.

Maenner, Matthew J. et al. "Prevalence and Characteristics of Autism Spectrum Disorder among Children Aged 8 Years—Autism and Developmental Disabilities Monitoring, 11 Sites, United States, 2018." Centers for Disease Control and Prevention, December 3, 2021. https://www.cdc.gov/mmwr/volumes/70/ss/ss7011a1.htm.

Manson-Smith, Alex. "Why Children with Autism Love Trains (and the 10 Best U.K. Rail Adventures for Families)." *The Telegraph*, June 19, 2018. https://www.tele graph.co.uk/travel/family-holidays/best-train-attractions-in-uk-and-why-kids-with -autism-like-trains/.

Meentemeyer, Blake. "Family Tees." USGA.org, November 10, 2016. https://www .usga.org/course-care/regional-updates/west-region/family-tees.html.

Minnaert, Lynn. "U.S. Family Travel Survey 2017," by the Jonathan M. Tisch Center for Hospitality and Tourism, NYU School of Professional Studies, and the Family Travel Association. January 2018. http://2ks7od1dgl654aorxx98q3qu-wpengine .netdna-ssl.com/wp-content/uploads/2018/01/FTA.NYU_.2017-Consumer-Fam ily-Travel-Survey.pdf.

Mshar, Alethea. "Special Needs Parenting: What Is a Social Story and How Can I Get One?" Shield Healthcare, September 6, 2018. http://www.shieldhealthcare.com /community/grow/2018/09/06/special-needs-parenting-what-is-a-social-story -and-how-can-i-get-one/.

Murrell, Jenna. "When You Should Pass on the Rental Car Seat." Safe Ride 4 Kids, October 6, 2016. https://saferide4kids.com/blog/why-you-should-pass-on-the -rental-car-seat/.

National Autism Association. "Autism Fact Sheet." 2021. https://nationalautismassocia tion.org/resources/autism-fact-sheet/.

National Institute of Mental Health. "Autism Spectrum Disorder Fact Sheet." 2021. https://www.nimh.nih.gov/health/topics/autism-spectrum-disorders-asd/index .shtml.

News 12 Staff. "Toms River Restaurant to Open Space Geared Toward Those with Autism." News 12 New Jersey, October 24, 2019. https://newjersey.news12.com /toms-river-restaurant-to-open-space-geared-toward-those-with-autism-41230451.

O'Brien, Kathleen. "Kids with Autism Take a Swing at Golf, Courtesy of PGA Pro Ernie Els." NJ.com, last updated March 29, 2019. https://www.nj.com/health fit/2015/06/golf_and_autism.html.

Operation Autism: A Resource Guide for Military Families. "By Car." 2021. https:// operationautism.org/transitions/travel/#1543352575441-1fe85f27-f5d1d42c-13af.

Operation Autism: A Resource Guide for Military Families. "Travel." 2021. https:// operationautism.org/transitions/travel/#1543437067753-19a08b6d-ff2f.

Perrachon, Juliet. "Just When You Think Traveling with Kids Can't Get Any Worse, It Does." Scarymommy.com, last modified June 24, 2021. https://www.scarymommy .com/travel-with-kids-horror-stories/.

Penny/blogger. "Using Public Restrooms with Your Child with Autism." Our Crazy Adventures in Autismland, 2021. https://ourcrazyadventuresinautismland.com /using-public-restrooms-with-children-with-autism/.

Philen, Tabatha. "Camping with Special Needs Kids." Meet Penny, July 16, 2016. https://www.meetpenny.com/2016/07/camping-special-needs-children-autism/.

Plane Finder. "Plane Spotting and Autism." Plane Finder, January 4, 2019. https:// planefinder.net/about/news/plane-spotting-and-autism/.

Ramah Darom. "Camp Yofi." 2021. https://www.ramahdarom.org/ramah-darom -jewish-summer-camp/summer-offerings/camp-yofi/.

Reed, J.R. "How Geocaching Works for the Autistic Me." The Good Men Proj-ect, February 23, 2019. https://goodmenproject.com/featured-content/geocaching -works-for-autistic-me-jrrd/.

Renzetti, Jackie. "For This Year's Summer Vacations, 97 Percent Will Travel by Car." Bring Me the News, June 29, 2020. https://bringmethenews.com/minnesota-news /for-this-years-summer-vacations-97-percent-will-travel-by-car.

Reynolds, Ann M., and Malow, Beth A. "Sleep and Autism Spectrum Disorders." National Jewish.org, 2011. https://www.nationaljewish.org/NJH/media/pdf/Melt zer%20References/Reynolds-(2011)-Sleep-and-autism-spectrum-disorders.pdf.

Riv's Toms River Hub. "Welcome to Chase's Friend Zone." 2021. https://rivstrhub .com/chases-friend-zone/.

Royal Caribbean International. "Autism Friendly Ships." 2021. https://www.royal caribbean.com/experience/accessible-cruising/autism-friendly-ships.

RTOR.org. "Borderline Personality Disorder." 2021. https://www.rtor.org/border line-personality-disorder.

Saeland, Jodi (Mountain Mama). "Skiing/Boarding Best Therapy for Those with Disabilities." Ski Utah, March 22, 2017. https://www.skiutah.com/blog/authors/jodi /for-those-with-disabilities-skiing/.

Sampson, Hannah. "8 Questions About Flying with Emotional Support Animals, Answered." *The Washington Post*, January 14, 2021. https://www.washingtonpost.com /travel/2021/01/14/emotional-support-animal-airlines-restrictions/.

Schuster, Blake. "PGA Tour's Billy Mayfair Says He Has Autism Spectrum Disorder." Bleacher Report, April 21, 2021. https://bleacherreport.com/articles/2940757-pga -tours-billy-mayfair-says-he-has-autism-spectrum-disorder.

Seeley, Amy. "Be Prepared: Carry an Autism Survival Kit." Autism Society of North Carolina, October 13, 2015. https://www.autismsociety-nc.org/be-prepared-carry -an-autism-survival-kit/.

Six Suitcase Travel. "How to Rent a Houseboat for a Fantastic Big Family Vacation." 2021. https://sixsuitcasetravel.com/blog/how-to-rent-a-houseboat-for-a-fantastic -big-family-vacation/.

Smith, Celeste. "7th Annual Autism Takes Flight Hosted at Wilmington International Airport." WWAYTV3, April 2, 2022. https://www.wwaytv3.com/7th-annual -autism-takes-flight-hosted-at-wilmington-international-airport.

Starling, Nick. "Dallas Airport Program Helps Children with Autism Get Real Life Airplane Experience." Kake.com, April 3, 2022. https://www.kake.com /story/46207534/dfw-airport-program-helps-children-with-autism-get-real-life-air plane-experience.

Surfside Beach, South Carolina. "Autism Friendly." 2021. https://www.surfsidebeach .org/autism-friendly.

Swift, Roxanna. "Roller Coaster Conqueror Logan Joiner, on the Autism Spectrum, Helps Others Overcome Their Fears." WCPO-9, Cincinnati, April 24, 2017. https://www.wcpo.com/news/insider/logan-joiner-addresses-his-fears-and-those -of-others-on-the-autism-spectrum-by-riding-and-reviewing-roller-coasters.

Tartakovsky, Margarita. "12 Travel Tips for People with Bipolar Disorder." PsychCentral, August 3, 2013. https://psychcentral.com/blog/12-travel-tips-for-people-with -bipolar-disorder.

The Daily Tee Blog. "10 Great Courses to Play with Your Kids (or Grandkids)." Golf Now.com, January 14, 2013. https://blog.golfnow.com/10-great-courses-to-play -with-your-kids-or-grandkids/.

Thibault, Nicole. "Are Autism Parents 'Cautious Travelers?'" Autismtravel.com, 2021. https://autismtravel.com/2018/02/02/are-autism-parents-cautious-travelers/.

Thibault, Nicole. "The Zoo Is a Perfect Outing for Children with Autism." Autism travel.com, 2021. https://autismtravel.com/2018/06/04/parent-perspective-the -zoo-is-a-perfect-outing-for-kids-with-autism/.

Thibault, Nicole. "Episode 124: Traveling with Kids on the Autism Spectrum." The Vacation Mavens.com, 2021. https://vacationmavens.com/124-autism-travel/.

Thompson, Dennis. "Autism Greatly Boosts Risk of Drowning." Health-Day News/WebMD, March 21, 2017. https://www.webmd.com/brain/autism/news/20170321/autism-greatly-boosts-kids-injury-risk-especially-for-drowning.

Tom, Kiana. "17 Tips You Must Know Before Renting a Houseboat." Kiana Tom, 2021. https://kiana.com/17-tips-must-know-before-renting-a-houseboat/.

Tourette Association of America. "Print an 'I Have TS' Card.'" 2021. https://tourette.org/about-tourette/overview/living-tourette-syndrome/print-ts-card/.

Tourette Association of America. "What Is Tourette." 2021. https://tourette.org/about-tourette/overview/what-is-tourette/.

Transportation Security Administration. "TSA PreCheck." 2021. https://www.tsa.gov/precheck.

Travel Community (Private Group by Scott's Cheap Flights). "What Are Some of the Biggest Problems You've Had When Traveling with Your Kids?" Facebook, December 12, 2020. https://www.facebook.com/groups/scftravelcommunity/permalink/1313555022340779.

Turner, Matt. "Stats: 65 Percent of U.S. Travelers Looking Forward to Traveling Again." Travel Agent Central, June 3, 2020. https://www.travelagentcentral.com/your-business/stats-65-percent-u-s-travelers-looking-forward-to-traveling-again.

U.S. Customs and Border Protection. "Global Entry." 2021. https://www.cbp.gov/travel/trusted-traveler-programs/global-entry.

U.S. Department of Health and Human Services/Health Resources and Services Administration/Maternal and Child Health Bureau. "The National Survey of Children with Special Health Care Needs." Last accessed October 20, 2021. https://mchb.hrsa.gov/chscn/pages/prevalence.htm.

U.S. Department of Transportation. "Traveling with a Disability." 2021. https://www.transportation.gov/individuals/aviation-consumer-protection/traveling-disability.

Valle, Marlene. "Traveling with Mental Illness: What Is It Really Like?" Deafinitely Wanderlust.com, May 2018. https://deafinitelywanderlust.com/2018/05/traveling-with-mental-health-illness-what-is-it-really-like/.

Visit Mesa. "Mesa, Arizona is the First-Ever Autism Certified City in the United States." 2021. https://www.visitmesa.com/mesaautismcertifiedtravel/.

Wigwam Holidays. "Glamping and Camping Tips for Families with Autistic Kids." 2018. https://blog.wigwamholidays.com/glamping-camping-tips-for-families-with-autistic-kids.

Wilderness Inquiry. "Families Together." 2021. https://www.wildernessinquiry.org/programs/accessible-family-vacations/.

Wilderness Inquiry. "Our Mission: Everyone Belongs in the Outdoors." 2021. https://www.wildernessinquiry.org/mission/.

Zizak, Angela. "An Autism Guide to Camping." Your Autism Guide, May 16, 2019. https://www.yourautismguide.com/2019/05/16/an-autism-guide-to-camping/.

Zizak, Angela. "National Park Access Pass for Autism." Your Autism Guide, May 10, 2019. https://www.yourautismguide.com/2019/05/10/national-park-access-pass-for-autism/.

INDEX

AAA Travel, 278, 285

AAW Gallery of Wood Art, 267

ACAA. *See* Airline Carrier Access Act

The Academy of Natural Sciences of Drexel University, 259

Accessible Destinations (podcast), 220

accommodation, 143; advocacy for, 217; for ASD, 21; at Beaches Resorts, 137; at destination, 42; diagnosis required by, 18; for flying, 42; Hardy on, 21–22, 210–11; with IAC, 186; at room, 102; for sleep, 125–26; Stull on, 128; technology in, 16–17; Zeihr on, 130. *See also* Autism on the Seas

Achieve Tahoe, 235

activity, 218

ADA. *See* Americans with Disabilities Act

Adaptive Ski Program at Jackson Hole Mountain Resort, 242

Adaptive Sport Center, 236

Adaptive Sports at Mount Snow, 241

Adaptive Sports at Snowbasin Resort, 240

Adaptive Sports Foundation at Windham Mountain Resort, 239

ADHD. *See* attention-deficit/hyperactivity disorder (ADHD), children with

adjoining rooms, 128–29, 144

Adlowitz, Jason, 72, 73–74, 85–86, 89, 165, 167–68, 198

adults, 4

Adventure Luxury Travel, 281–82

Adventure Ocean Youth Program, 115

Adventures in Golf, 283–84

advocacy, 209, 211, 217, 219–21

AF. *See* autism-friendly

Airbnb, 37, 123, 126–27, 135, 229

Airline Carrier Access Act (ACAA), 219

airplane, 65–68, 70–71, 207–8

airports, 67–68, 93, 210

Alaska Museum of Science and Nature, 257

Alibates Flint Quarries Tours, 257

All Aboard America! Bus Tours, 170

All About Travel, LLC, 287

Alligator and Wildlife Discovery Center, 172

American Clock & Watch Museum, 249

American Computer & Robotics Museum, 251

American Dream Parks, 171

American Museum of Natural History, 178, 253, 257

American Society of Travel Agents (ASTA), 215–16

Americans with Disabilities Act (ADA), 209, 219

America the Beautiful National Parks and Federal Recreation Lands Access Pass, 151–52

Amtrak, 100, 207, 247

amusement parks, 164–66

Andy Warhol Museum, 182
Antell, Judy, 67–68
Antique Boat Museum, 267
AotS. *See* Autism on the Seas
aquarium, 33–34
Aquatica Orlando Water Park, 173
Arizona Goat Yoga, 170
Arizona Museum of Natural History, 170
Arky, Beth, 50
Arky, Carol, 21
Ascendigo, 236
ASD. *See* autism spectrum disorder (ASD), children with
Association of Children's Museum, 163
Association of Science and Technology Centers, 258
ASTA. *See* American Society of Travel Agents
Atlantis Resorts Waterpark, 166
Atlas Obscura, 269
attention-deficit/hyperactivity disorder (ADHD), children with: ASD compared with, 15–16; behavior of, 11; flying with, 37; mental health organizations for, 276; travel with, 9, 29
Attwood, Tony, 24, 53–54, 130–31, 290–91
autism, children with: ADHD compared with, 15–16; adults with, 4; in airports, 67–68; Antell on, 67–68; awareness for, 8; at destination, 61; with disappointment, 105; internationality understanding, 56; mental health organizations for, 275–76; museum prioritizing, 162–63; National Autism Association defining, 14; noise with, 48; roller coasters interesting, 164; sensory issues accompanying, 14; stress with, 44; at zoo, 30. *See also* Autism on the Seas
Autism Adventure Guide, 31–32, 169, 194
The Autism Channel, 114
Autism Double-Checked, 32

autism-friendly (AF), 33, 39, 113–15; Beaches Resorts as, 137; destination as, 227; dude ranch as, 140–42; hotel as, 127; Leafy Fields as, 147–48; museum as, 163; New York City as, 229; resort as, 132–36; Royal Caribbean as, 113–14; scuba as, 242–45; venues as, 168–85
Autism in Museums, 163
Autism Inspiration, 50
Autism on the Seas (AotS), 120, 187; Cruise Assistance Package provided by, 117; cruises offered by, 114; financial assistance offered by, 117; McKerley on, 110; Royal Caribbean collaborating with, 116
Autism Parenting Magazine, 14–15, 243
Autism Speaks, 74–75, 218
autism spectrum disorder (ASD), children with: accommodation for, 21; awareness of, 227; behavior of, 8; Bennett on, 99; bipolar disorder mistaken for, 9–10; boarding expedited for, 110; in crowds, 159–60; *DSM* classifying, 14–15; familiarity in, 131; in family, 18; geocaching enjoyed by, 153; golf aiding, 230–34; location considering, 128; with motion sickness, 83; neurotypicality differentiated from, 7–8, 13–14, 23–24; packing for, 46–48; pre-boarding for, 69; preparation with, 57; rentals suiting, 126; routine with, 37; swimming with, 242; toileting with, 211; trains fascinating, 100, 103–6, 163–64; travel impacted by, 5–6, 13–21, 29, 51–52, 59–60, 142–43; vacation with, 4, 54; water enjoyed by, 243. *See also* autism, children with; autism-friendly
Autism Takes Flight, 67
Autism Travel, 31, 124, 215
Autism Travel Directory, 132, 169
Autism Village, 163, 194, 200
awareness, 8, 227

BAA Training, 248
Badlands Dinosaur Museum, 253
Baeszler, Dawn, 46, 218–19, 226, 278
Baltimore Streetcar Museum, 264
Barber Vintage Motorsports Museum, 262
Barclay, Dawn M., 3
Bart Adaptive Sports Center at Bromley Mountain, 241
Bayer Insectarium at Saint Louis Zoo, 256
Beaches-Ocho Rios, 204
Beaches Resorts, 58, 136–38
Beaches Resorts, Negril, 136
Beaches Resorts, Ocho Rios, 136
Beaches' Special Services Team, 138
Bear Mountain, 235
Beaver Creek Resort, 236
bedtime, 131
behavior, 8, 11, 79, 102
benefit, 17, 19, 38, 56, 188–89, 228
Bennett, Amanda, 99
Berkie Travel, 278
Berkman, Leslie, 278
Berkshire Scenic Railway Museum, 260
Bertuccio, Lisa, 8, 23, 49, 56; on airplane, 65–66; on boats, 246–47; on cabin, 111; on camping, 149; on DPNA, 70; go-to bag recommended by, 197; insurance emphasizing, 205–6; on motion sickness, 84; preparation considered by, 219; on tours, 187; trigger discussed by, 82–83, 129, 160; with Vacation Your World, 278–79
BGB Travel and Leisure, 279–80
Big Cedar Lodge, 233
Bigley, Michele, 13
Bily Clocks Museum, 249
bipolar disorder, 9–10, 214–15, 277
Blackhawk Museum, 262
Bluegreen Resorts, 32–33
Blumhoefer, Nate, 243
BluWater Scuba, 245
boarding, airplane, 110
boats, 246–47

B&O Railroad Museum, 260
borderline personality disorder (BPD), 215
Boston's Children's Museum, 175
Bowlero Lanes and Lounge, 176
BPD. *See* borderline personality disorder
breaks, in time, 22, 25, 96–97, 190
Bright Horizons Diving, 244
The Broadmoor, 232
Brooklyn Children's Museum, 179
Brooklyn Junior Autistic Golfers Academy, 231
Brown, Chandra Nims, 54, 69, 103, 210, 217, 279
The Bug and Reptile Museum, 256
Burger, Betty J., 48–49, 55, 64, 101, 107, 279–80; on familiarity, 147
bus, 90
Butler Institute of American Art, 180
Butterfly Wonderland, 170
Buxton, Matt, 153

C2C Escapes Travel, 285
cabin, 111, 147
CACs. *See* Certified Autism Centers
California Auto Museum, 262
California Science Center, 258
California State Railroad Museum, 260
campgrounds, 148–51
camping: Bertuccio on, 149; in cabin, 147; geocaching during, 152–53; Query-Fiss on, 226–27; Romans discussing, 148–49; sensory issues while, 147; sleep with, 145; structure during, 149–50; in tent, 145–46, 149; with Wilderness Inquiry, 188–89; Zizak on, 147
Camping with Special Needs Kids (Philen), 149–50
Camp Yofi, 148
CAN. *See* Champion Autism Network
Canadian Grain Elevator Discovery Centre, 255
Cannon Mountain Adaptive Sports Partners of the North Country, 238

car, 90–91
CARD. *See* Center for Autism and Related Disorders
Care.com, 203
caregiver, 211
Carowinds Theme Park, 180
car seat, 71–72, 91
CATPs. *See* Certified Autism Travel Professionals
Cedar Point Amusement Park, 180
Center for Art in Wood, 268
Center for Autism and Related Disorders (CARD), 19, 127
Centers for Disease Control and Prevention, 4
The Center for Science and Industry (COSI), 259
Central Park Zoo/Wildlife Conservation Society, 179
Certified Autism Centers (CACs), 18, 32, 168–85; Advanced, 31, 57–58, 132–37; Lake Havasu City as, 227; Marshall on, 83; Mesa as, 225, 227; in Orlando, Florida, 228–29; in Philadelphia, 229; reentry policy of, 206; Salter on, 69; Tekin on, 206; in 2021, designated as, 132, 169
Certified Autism Travel Professionals (CATPs), 66, 108, 120, 277–91; Autism Travel locating, 215; cruise offering, 116; destination suggested by, 226; hotel with, 123; IBCCES creating, 124; St. Pierre pursuing, 18–19. *See also* Haraty, Jonathan; Hardy, Jennifer; Marshall, Sarah; Romans, Angela; Salter, Sarah; Speid, Marian; Stull, Belinda S.; Wallace, Nicole Graham; Wlodarski, Starr; Woodbury, Tara
Challenge Alaska Adaptive Ski and Snowboard School, 235
Challenge Aspen at Snowmass, 236
Challenged Athletes of West Virginia, 242

Chambliss, Kristen, 21, 48, 142–43, 149; on Beaches-Ocho Rios, 204; on enthusiasm, 92; on rudeness, 81; sensory issues discussed by, 157–58; on Social Stories, 66
Champion Autism Network (CAN), 32, 228
Charman, Tony, 164
The Chelsea Hotel, Toronto, 135
Cheshire Children's Museum, 177
Chicago Children's Museum at Navy Pier, 174
Chick-fil-A College Football Hall of Fame, 173
children: on airplane, 65; culture experienced by, 38; destination selected by, 43–44; familiarity comforting, 47; plan assisting, 22–23, 41; preparation involving, 43–44; travel with, 3–4, 7–12; vacation exciting, 42. *See also* attention-deficit/hyperactivity disorder (ADHD), children with; autism, children with; neurodiversity, children with; neurotypicality, children with
The Children's Museum of Atlanta, 174
Children's Museum of Houston, 183
Children's Museum of Indianapolis, 174
Children's Museum of Manhattan, 179
Children's Museum of New Hampshire, 177
Children's Museum of Pittsburgh, 182
Children's Museum of the Arts, 179
Chu, Sandra, 41, 57, 81–82, 84, 102, 280; on destination, 159
Circuit of the Americas, 266
Clark, Pilar, 5
Clemens, Erin, 153
CLIA. *See* Cruise Lines International Association
Coffee Creek Ranch, 141
Colorado, 228, 260
Colorado Railroad Museum, 260
Comfort Suites on Paradise Island, 166

Common Ground Outdoor Adventures, 240
communication, 61
community, 104, 153
Computer History Museum, 250
Computer Museum of America, 250
computers, 250–52
Concannon, Terence, 280–81
Connecticut Trolley Museum, 264
COSI. *See* The Center for Science and Industry
Courtyard Phoenix Mesa, 133
Cousins, Lucy, 35
COVID-19 pandemic, 4, 125
Covington, Taylor, 43–48, 82, 94, 160; adjoining rooms recommended by, 128–29; on car seat, 91; on preboarding, 78; on routine, 95; sensory issues discussed by, 197–98; on Social Stories, 109; on suppliers, 217; toileting discussed by, 218
Crayola Experience, 176
Creative Discovery Museum, 182
Crites, Steven, 164
crowds, 64, 159–60, 162, 229
cruise, 54, 61, 119–20; CATPs offered by, 116; dining on, 111–12; Hardy on, 85, 108, 109; Marshall on, 108; modification for, 113; with older neurodiverse children, 210–11; Roberts on, 108; Social Stories for, 115; with special needs program, 115–16; staff assisting with, 114, 116–17; trial run allowed by, 109; Woodbury recommending, 42, 112. *See also* Autism on the Seas; *specific cruise*
Cruise America, 151
Cruise Assistance Package, 117
Cruise Lines International Association (CLIA), 215
Cruise Planners, 205, 279, 282–83
Cruising the World, 59
culture, 38–39, 53
Custom Travel Services, Inc., 283

Dallas Museum of Art, 183
Daniels, Natasha, 47, 131
Da Re, Rebecca, 101
DAS. *See* Disability Access Service
Daytona International Speedway, 265
Delta Hotel Phoenix Mesa, 133
Denver Museum of Nature & Science, 252
Department of Transportation, 206–9
destination, 12; accommodation at, 42; as AF, 227; autism at, 61; CATPs suggesting, 226; children selecting, 43–44; Chu on, 159; familiarity lacked by, 56; family choosing, 20; picture book on, 30–31; Polak considering, 216; rentals at, 190; train station as, 101; videos previewing, 61; weather at, 216, 222. *See also specific destination*
Destination Sitters, 203–4
Details by Bill LLC, 284
diagnosis, 18
Diagnostic and Statistical Manual of Mental Disorders (DSM), 14–15
dialectical behavior therapy, 215
Diaz, Natalia, 36
dining: on cruise, 111–12; familiarity smoothing, 200; Heckman-Caliciotti on, 194–95; Hutten on, 198; Johnson, C., on, 195; Sokol on, 130; Stull on, 112; Woodbury considering, 193; Zeihr on, 79. *See also* restaurant
Disability Access Service (DAS), 159, 167–68
Disabled Passenger with Intellectual or Developmental Disability Needing Assistance (DPNA), 70–71
Disabled Sports Eastern Sierra, 235
disappointment, 105
Discovery Cove Theme Park, 173
Discovery Place, 259
Disney, 160, 166–68, 198
Disney Access Pass, 167
Disney Parks and Resorts, 169
Disney World, 159

disorders: bipolar, 214–15, 277; mood
    and neurological, 213–15; Tourette
    Syndrome as, 213, 277; trigger
    with, 214. *See also* attention-deficit/
    hyperactivity disorder
distraction, 144; packing for, 105;
    Romans prioritizing, 130; in RV,
    155; Sokol emphasizing, 196; Speid
    on, 79; Woodbury on, 84, 103
Dive Georgia, 244
Diveheart, 242–44
Dive Paradise, 244
Dixie Dude Ranch, 142
Dogwood Canyon Nature Park, 177
Dolcine, Delight, 46–47, 53, 82, 89, 128,
    132; insurance considered by, 217; on
    restaurant, 193; on tours, 162
Dollywood, 183
Don Garlits Museum of Drag Racing,
    262
Don's Bay Getaway Airbnb, 135
Dorney Park and Whitewater Kingdom,
    181
Doubletree by Hilton Hotel Orlando,
    133
DPNA. *See* Disabled Passenger with
    Intellectual or Developmental
    Disability Needing Assistance
DREAM Adaptive Recreation, 237
drowning, 149
Dr. Phillips Center for the Performing
    Arts, 173
*DSM. See Diagnostic and Statistical Manual
    of Mental Disorders*
dude ranch, 140–42, 144
DuPage Children's Museum, 174

Eagle Mount at Bridger Bowl, 238
Edaville Family Theme Park, 175
#EGameON (program), 230
Electric City Trolley Museum, 264
The Elevator Museum Inc., 254
elevaTOURS Elevator Museum, 255
Elite Events & Tickets, 284–85
Elliott, Jim, 242–43

Elmwood Park Zoo, 181
elopement, 71, 72, 160
emotional service animals (ESAs), 206–7,
    221
The Enchanted Traveler, 280
Engstrom, Kellie, 96–97, 132, 165, 217,
    219
enthusiasm, 92
EnVision Travel, 289
equine therapy, 140, 218, 222
Erik's Ranch, 141
ESAs. *See* emotional service animals
Escape into Travel, 290
Exceptional Vacations, Inc., 213
expert, 290–91
Exploratorium, 258
Explore & More, 177
Extra Hands, 160

Fairfield Inn and Suites by Marriott
    Orlando at SeaWorld, 134
familiarity: in ASD, 131; Burger on, 147;
    children comforted by, 47; destination
    lacking, 56; dining smoothed by, 200;
    Jones prioritizing, 112–13; Keiper
    emphasizing, 112–13; Romans on, 56
Families Integrating Together, 188
family, 3–6, 16–20, 137–38
Family Museum, 175
Family Travel Association (FTA), 4,
    17–18, 20
The Family Travel Handbook, 16
Family Trip, 36
Fast, Julie A., 214–15
fast passes, 165
The Field Museum, 253
financial assistance, 117
Fineman, Scott, 212–13
flexibility, 216
flight, 68–69, 71, 83
flying, 3, 23, 37, 51, 206, 248–49;
    accommodation for, 42
Fort Myers Mighty Mussels, 172
Fort Wilderness at Walt Disney World,
    151

Fort Worth Zoo, 183
4 All Senses, 59
Four Points by Sheraton at Phoenix
    Mesa Gateway Airport, 133
Four Rivers Environmental Education
    Center, 174
Francus, Margalit, 139
Franklin Institute, 259
Franklyn D. Resort & Spa, 136
The Friends Experience, 178
Frontier City, 180
Frontier Travel Camp, 212–13
Frosch Travel, 285–86
FTA. *See* Family Travel Association
functioning level, 212–13
Funtherapy Travel, 281

Gabrieli, John, 245–46
Gardner, Alan, 148
gate agent, at airport, 77
Gateway to Adventure, 188
geocaching, 152–53
Geocaching.com, 153
Georgia Aquarium, 173
Geronimo Trail Guest Ranch, 141
Gilbreath, Sandy, 36
Glass Slipper Concierge, 284
Glazer Museum, 173
Go Lake Havasu, 280–81
Gold Coast Scuba, 244
golf, 230–34
Golf Digest, 231
The Golisano Children's Museum, 173
The Good Men Project, 152
Gotham Divers, 245
go-to bag, 22; Bertuccio recommending,
    197; Romans suggesting, 44;
    sensory issues aided by, 165; Speid
    recommending, 71, 94, 111
The GRAMMY Museum L.A. Live, 171
Grand Palladium Bavaro Suites Resort &
    Spa, 136
Granite State Adaptive at King Pine at
    Purity Spring Resort, 238
Gray, Carie-Louise, 101–2

Gray, Carol, 50
Great Sand Dunes National Park and
    Preserve, 172
Great Wolf Lodge, 169
Grovewood Gallery and Museum, 268
Guest Services, 167
The Guided Tour, 213
Gunstock Mountain Resort Lakes
    Region Disabled Sports, 238

Hall, Markeisha, 210, 281
Hamed, Hend M., 158
Hammer Travel, 213
Hands On Children's Museum, 184
Haraty, Jonathan, 41, 54; of Adventure
    Luxury Travel, 281–82; on bus, 90;
    on hotel, 130; on houseboat, 110;
    insurance recommended by, 205; on
    location, 126–27; meltdown avoided
    by, 166; on older neurodiverse
    children, 210; preparation discussed
    by, 147; on tours, 161; on trigger, 67,
    99, 145–46, 165
Hardy, Jennifer, 29, 42–44, 81, 97, 218;
    on accommodation, 21–22, 210–11;
    communication emphasized by,
    61; on cruise, 85, 108, 109; Cruise
    Planners owned by, 282–83; on
    DAS, 168; on elopement, 160; on
    mealtime, 103; plan discussed by,
    71–72, 215–16; on restaurant, 112,
    196; RV suggested by, 146, 148;
    sitters discussed by, 205; Social Stories
    suggested by, 48–49; on SSR, 70;
    on stimuli, 123; trains discussed by,
    246; on trigger, 216; on TSA Cares,
    75–77; TSA PreCheck recommended
    by, 210; water considered by, 246
Harstad, Elizabeth, 15–16
Heckman-Caliciotti, Jodi, 66, 194–95
Henry Ford Museum of American
    Innovation, 263
Hensley, Aaron, 230
Hershey's Chocolate World Attraction,
    181

Herskovitz, Janeen, 137, 291
Heubish, Michelle, 9, 10–12
Higher Ground, 237
High Museum, 267
Holden, Nell, 188–89
Holiday World and Splashin' Safari Park, 174
Hong, Jen, 230
Horizons Skiing, 237
"Horseback Riding for Kids with Autism" (Kilgore), 140
The Horse Boy (Isaacson), 140
Horseshoe Bay Resort, 234
hotel, 11; as AF, 127; with CAPT, 123; checking into, 130; noise in, 48; older neurodiverse children in, 210; rentals compared with, 126–27
Hotel Port Aux Basques, 136
houseboat, 110, 118–19
Houston Museum of Natural Science, 183, 256
Hudson Valley Maritime Museum, 268
Hutten, Mark, 165–66, 197–98

IAC. *See* International Board of Credentialing and Continuing Education Standards Accessibility Card
IBCCES. *See* International Board of Credentialing and Continuing Education Standards
Idaho Potato Museum, 269
I.D.E.A Museum, 170
Ignite Adaptive Sports at Eldora Ski Resort, 236
Illinois Railway Museum, 260
incontinence, 211
Indianapolis Motor Speedway, 265
Indianapolis Motor Speedway Museum, 263
industry, 219–20
Insectarium at Audubon Nature Institute, 256
insurance, 205–6, 217, 221
Intel Museum, 250

interests: aviation as, 248–49; clocks and watches as, 249; computers as, 250–52; dinosaurs as, 252–54; elevators as, 254–55; insects as, 255–56; passion created by, 271; as quirky, 269–71; racing as, 265–66; rocks as, 257; science as, 258–60; transportation as, 260–66; vacation around, 245–71; videogames as, 250–52; woodworking as, 266–69. *See also* trains
International Board of Credentialing and Continuing Education Standards (IBCCES), 18, 31–34, 124, 132–36
International Board of Credentialing and Continuing Education Standards Accessibility Card (IAC), 81–82, 125, 186
International Cryptozoology Museum, 270
internationality, 54, 56
International Spy Museum, 269
International UFO Museum and Research Center, 270
Intrepid Sea, Air, and Space Museum, 178, 249
Iowa 80 Trucking Museum, 261
Isaacson, Rupert, 140
Island H2O Live! Water Park, 172
itineraries, 225–27
"It's Cool to Fly American," 67

Jaworskyj, Georgiann, 127, 283
Jenss, Rainer, 13, 17, 20–21
JetBlue, 67
The Jewish Museum, 179
J.H. Hawes (Grain) Elevator Museum, 254
Johnny Morris' Wonders of Wildlife National Museum & Aquarium, 177
Johnson, Carl W., 55, 83, 107, 209, 232–34; of Adventures in Golf, 283–84; on bus, 90; on dining, 195; motion sickness discussed by, 85; on rehearsal, 68; rental discussed by, 126; on research, 41–42; on RV, 146; on trains, 90, 100
Johnson, Dan, 243

Johnson, William, 43, 284
Jones, Jesemine, 111, 112–13
Jordan Schnitzer Museum of Art, 180
Jorgensen, Theresa, 118–19
journal, 209
Juve Travel Peru, 58–59

Kansas Aviation Museum, 248
The Kansas Barbed Wire Museum, 269
Kay, Alice, 243
Keiper, Ida, 111, 112–13
Kennett Classic Computer Museum, 251
Kennywood Theme and Amusement
   Park, 182
Keystone Adaptive Center at Keystone
   Ski Resort, 236
Kiawah Island Golf Resort, 233
KidsQuest Children's Museum, 184
KidzMondo Doha, 186
Kilgore, Gene, 140, 142
Kings Dominion and Soak City, 184
Kings Island Amusement Park, 180
kitchen, 129
Knoebels Amusement Resort, 181
Knott's Berry Farm, 171

Lake Havasu City, Arizona, 227
Lane Motor Museum, 264
Large Scale Systems Museum, 251
Lead with Love Training Company, 32
Leafy Fields, 147–48
Lee, Cory, 36 ·
LEGOLAND, 169
LEGOLAND Discovery Center
   Philadelphia, 182
LeMay-America's Car Museum, 264
Let's Explore with Cor Cor (Lee and
   Gilbreath), 36
Liberty Mountain Resort, 240
Liberty Science Center, 253, 259
Lime Rock Park, 265
Lindsey, Alan, 234
Little Traveler Board Book Set, 35
Littman, Ellen B., 6–8, 22, 80, 128, 157,
   291; on train station, 101

Living Computers: Museum & Labs, 252
location, 126–28, 144, 151
locks, 147
Long, Alyson, 7
The Long Ride Home (Isaacson), 140
Los Angeles County Fire Museum, 261
Loves Park Scuba, 245
Ludwig, Sarah, 189–90

Mack Trucks Historical Museum, 262
MacLachlan Woodworking Museum,
   269
Madge, Claire, 163
Madlenka (Sis), 31, 36
Magical Storybook Travels, 288–89
Maine Adaptive at Sunday River Resort,
   237
Maisy Goes on a Plane (Cousins), 35
Mall of America, 176, 229
Marciniak, Dana, 132, 199–200
Marian Speid Dream Vacations, 287
MarineLand Canada, 185
Marshall, Sarah, 19–20, 48, 54, 85, 128;
   on CACs, 83; caregiver discussed by,
   211; on cruise, 108; on Disney, 160;
   of TravelAble, 284
De Mauro, Mike, 220–21
Mayfair, Billy, 230
May Natural History Museum, 255
Maziarz, Cathi Marcheskie, 284
McGee, William J., 69
McKerley, Olivia, 92, 110, 125–26, 157,
   219, 284–85
mealtime, 103, 196
Medicine Bow Lodge, 142
meltdown, 166, 189, 201, 206
mental health organizations, 275–77
Mermaid Travel, 288
Mesa, Arizona, 220, 225, 227
Mesa Arts Center, 170
The Metropolitan Museum of Art, 178
Miami Children's Museum, 172
Michigan's Adventure Theme Park, 176
Microsoft Visitor Center, 251
Mike's Mission Facebook page, 220

Military Technology Museum of New Jersey, 261
Minneapolis Institute of Arts, 267
Minnesota, 229
Mint Museum, 268
Modern Travel Professionals, 287–88
modification, 21–25, 38, 113
Molly and the Magic Suitcase Series (Oler), 35
Morgan's Wonderland Theme Park, 184
motion sickness, 84–85, 87
Mount Sunapee Resort's New England Healing Sports Association, 239
Mount Vernon, 184
Mshar, Alethea, 50
Mt Attitash Mountain Resort, 238
Murrell, Jenna, 91
museum: as AF, 163; autism prioritized by, 162–63; horological, 249; on transit, 163–64; on transportation, 260–64. *See also specific museum*
Museum of Contemporary Art, 172
Museum of Discovery, 258
Museum of Fine Arts, 175
Museum of Fine Arts Boston, 267
Museum of Modern Art, 178
Museum of Science, 253, 259
Museum of Science and Industry, 258
Museum of Technology at Middle Georgia State University, 250
Museum of the American Revolution, 181
muster drill, 113
Mutter Museum at the College of Physicians of Philadelphia, 270
My ASD Child (Hutten), 165, 197
Myrtle Beach, South Carolina, 228

NARM. *See* North American Reciprocal Museum
National Ability Center, 184
National Ability Center at Park City, 240
National Air and Space Museum, 258
National Autism Association, 14
National Autistic Society, 147–48, 162

National Automobile and Truck Museum, 263
National Automobile Museum, 263
National Building Granite Quarries Association, 257
National Building Museum, 252
National Child Identification Program, 45
The National Construction Equipment Museum, 252
National Geographic Little Kids First Big Book of the World (Carney), 36
National Museum of Funeral History, 270
National Museum of Military Vehicles, 262
National Museum of the U.S. Air Force, 249
National Mustard Museum, 245, 271
National Sport Center for the Disabled, 237
National Watch and Clock Museum, 249
neurodiversity, children with: campgrounds for, 148; drowning and, 149; insurance suiting, 205; neurotypicality contrasted with, 96, 190, 200; noise impacting, 165; sunflower lanyards identifying, 104–5; surprise with, 48; on vacation, 85–86. *See also* attention-deficit/hyperactivity disorder (ADHD), children with; autism, children with; autism spectrum disorder (ASD), children with
neurotypicality, children with: ASD differentiated from, 7–8, 13–14, 23–24; bipolar disorder demonstrated with, 10; in family, 16–20; neurodiversity contrasted with, 96, 190, 200
Nevada Northern Railway Museum, 261
New Directions Travel, 213
New England Disabled Sports, 238
New Orleans Pharmacy Museum, 269
New York City, New York, 229

New York Transit Museum, 178
The New York Transit Museum, 164
Nickelodeon Hotels & Resorts, Punta
  Cana, 136
Nickelodeon Universe, 176
noise: with autism, 47, 48; in car,
  91; in crowds, 229; in hotel, 47;
  neurodiversity impacted by, 165;
  Romans on, 111; Stull discussing,
  196–97; on trains, 102
North American Reciprocal Museum
  (NARM) Association, 247
North Carolina Granite Corporation,
  257
North Carolina Maritime Museum at
  Southport, 180
North Carolina Museum of Natural
  Sciences, 253
Northstar at Tahoe, 235
novelty, 149, 155
NY Transit Museum, 261

Oakland Athletics Spring Training
  Home, 170
Odysea Aquarium, 170
older neurodiverse children, 208–13
Oler, Chris, 35
Omni Houston, 135
Omni Mount Washington Resort, 233
One-on-One Beaches Buddy, 138
online review, 42
onlookers, 80–83
O. Orkin Insect Zoo, 255
Operation Autism, 68, 70–71, 95–96
Option Institute, 291
Orlando, Florida, 228–29, 232
Outdoors for All, 241
overnight, 30
overstimulation, 159–61, 165–66
Owens, Melissa, 36

Pacific Science Center, 185, 256
packing, 46–48, 52, 71–72, 94, 105
parent, 3, 17–18, 22, 209, 227
Passenger Service System (PSS), 209

passion: interests creating, 271; schedule
  relieved with, 31; vacation built from,
  225; venues feeding, 198, 247
pat-down, by security, 74
Path International Member Center, 218
Pennsylvania Ctr for Adaptive Sports,
  239
Peppa Pig (fictional character), 36, 149
Perot Museum of Nature and Science,
  254
Perrachon, Juliet, 3
Petersen Auto Museum, 262
Philadelphia, Pennsylvania, 229
Philadelphia Insectarium and Butterfly
  Pavilion, 256
Philen, Tabitha, 149–50
pickiness, 5, 37, 48, 71, 129, 197–98;
  Salter on, 131
picture book, 30–31, 35–36
The Pima Air and Space Museum, 248
Pincomb, Myron, 138
plan, 195; alternative, 45; for breaks,
  190; children assisted by, 22–23,
  41; Engstrom discussing, 219;
  Hardy discussing, 71–72, 215–16;
  internationality impacting, 56; with
  older neurodiverse child, 208–9; at
  restaurant, 211; with restaurant, 196;
  for road trip, 92; Romans discussing,
  44, 70; for sensory issues, 44, 49;
  Sokol on, 56; for travel, 7; Wlodarski
  on, 158. *See also* routine; schedule
Plane Finder Live Flight Tracker app,
  248
Platinum Protocols of Cleanliness, 138
Please Touch Museum, 181
Polak, Suzanne, 216, 285
pre-boarding, 69, 78
preparation: with ASD, 57; Bertuccio
  considering, 219; children involved
  in, 43–44; Chu on, 102; Haraty
  discussing, 147; for motion sickness,
  85; for road trip, 91; Romans
  considering, 216; routine maintained
  by, 24–25; Salter suggesting, 54;

Sokol discussing, 47; travel requiring, 92–93; for vacation, 154. *See also* plan previewing, 66–68, 143

PSD. *See* psychiatric service dog

PSS. *See* Passenger Service System psychiatric service dog (PSD), 207–8, 221

Puzzle Peace Counseling, LLC, 291

Query-Fiss, Adrienne, 127, 145, 154–59, 167, 226–27

Racine Art Museum, 268
rail travel, 163–64
Rainbow Reef, 244
Ramah Darom organization, 148
Ramon's Village Resort, 57–58, 135
The Ranch at Rock Creek, 141
Rancho Las Cascadas, 142
Reading Public Museum, 182
recreational vehicle (RV), 146, 148, 155
Reed, J. R., 152–53
rehearsal, 49, 68, 158
Rejuvia Travel, 286
rentals, 37, 140; ASD suited to, 126; at destination, 190; foreign, 190; hotel compared with, 126–27; Johnson, C., discussing, 126; research evaluating, 139; for vacation, 20
Renwick Gallery of the Smithsonian American Art Museum, 267
research, 41–43, 52, 139
resort, 57–58; as AF, 132–36; Herskovitz on, 137; Jaworskyj on, 127; older neurodiverse children in, 210; self-service, 127
restaurant, 11–12; Dolcine on, 193; Hardy on, 112, 196; Marciniak discussing, 199–200; menu of, 196; plan at, 211; plan with, 196; Romans discussing, 194–95; waiting at, 199–200; Wallace considering, 196; Woodbury on, 197; Zeihr considering, 194

rest stop, 95
Revere, Brenda, 17–18, 47, 109, 160, 209, 285; on safety, 149
Rhode Island Computer Museum, 251
Richard Scarry's Busy, Busy World (Scarry), 36
rights, 208–9
Ripley's Aquarium of Canada, 185
Ripley's Aquarium of Myrtle Beach, 182
Ripley's Aquarium of the Smokies, 183
Ripley's Believe It or Not!, 176
Riverbound Sports Paddle Co., 245
Riv's Toms River Hub, 194
Road America, 266
road trip, 91–92, 95–96
Roberts, Holly, 55, 69, 89, 108, 128, 285–86
Rock of Ages Visitors Center, 257
roller coasters, 164
Romans, Angela, 91–92, 94–95, 110, 112, 159, 161; camping discussed by, 148–49; distraction prioritized by, 130; on familiarity, 56; go-to bag suggested by, 44; on noise, 111; plan discussed by, 44, 70; preparation considered by, 216; with Rejuvia Travel, 286; restaurant discussed by, 194–95; routine prioritized by, 8; on trains, 103; on transfers, 93
room, 102–3, 106
routine, 118; with ASD, 37; Covington on, 95; meltdown without, 189; preparation maintaining, 24–25; Romans prioritizing, 8; transition in, 8; travel disrupting, 5, 7, 37
Royal Caribbean, 113–16
rudeness, 81
Rudge, Jason, 67–68
RV. *See* recreational vehicle

Saeland, Jodi, 234
safe foods, 197–98, 201
Safe Ride 4 Kids (Murrell), 91
safety, 45–46, 52, 53, 139–40, 155, 191; of car seat, 91; Engstrom on,

165; Revere on, 149; Salter on, 161; Woodbury on, 72, 108
Saguaro Lake Guest Ranch, 133
Salt and Pepper Shaker Museum, 270
Salter, Sarah, 44, 49, 80, 94–95, 132, 225; on advocacy, 211; on behavior, 102; on CATP, 69; incontinence considered by, 211; on muster drill, 113; on pickiness, 131; preparation suggested by, 54; on RV, 146; on safety, 161; with Sarah Travels Happy, 286–87; on Social Stories, 158
San Diego Air & Space Museum, 248
San Diego Zoo, 171
Santa Barbara Zoo, 171
Sarah Travels Happy, 286–87
Sawgrass Marriott Golf Resort and Spa, 134
Scarry, Richard, 36
schedule, 44, 98; breaks in, 22, 25; with flying, 23; for mealtime, 196; of overnight, 30; passion relieving, 31; security increased with, 21
Schlitterbahn Galveston, 183
Schlitterbahn New Braunfels, 184
school, 12
Science Museum of Minnesota, 259
scuba, 242–45
Scubagym at Scuba Quest Orlando, 244
Sea Island Resort, 232–33
Seattle Children's Museum, 185
SeaWorld, 169
SeaWorld Orlando, 173
SeaWorld San Diego, 171
security, airport, 21, 73–74, 118–19
sensory issues, 25, 95, 100, 140, 191; autism accompanied by, 14; while camping, 147; Chambliss discussing, 157–58; Covington discussing, 197–98; go-to bag aiding, 165; plan for, 44, 49; at theme park, 165; Woodbury on, 160. *See also* noise; trigger
Sensory Stimulation Guide, 138

Service Animal Transportation Form, 208
Sesame Place Philadelphia, 181
"17 Tips You Must Know Before Renting a Houseboat" (Tom), 119
The Shaker Museum, 268
Sheard, Andrew, 148
Sheard, Dannie, 148
Shedd Aquarium, 174
Sheraton Mesa Hotel at Wrigleyville West, 133
sightseeing, 161–62, 191
Sikora, Darryn, 153
Simeone Foundation Automobile Museum, 263
Sis, Peter, 31, 36
sitters, 204–5, 221
Six Flags Darien Lake, 178
Six Flags La Ronde, 185
Ski Apache Adaptive Sports, 239
skiing, 234–42
Ski Santa Fe and Albuquerque Adaptive Sports Program, 239
sleep, 125–26, 130–31, 144–45, 214–15
Smithsonian Institution, 162–63, 172
Smithsonian National Museum of American History, 250
Smithsonian National Museum of Natural History, 252
Smithsonian's National Air and Space Museum, 248
Smoky Mountain Trains Museum, 261
Smuggler's Notch, 135, 241
Soaring Eagle Waterpark and Hotel, 175
Sobbell, Michael, 116
social media, 104
Social Stories, 22, 50–51, 86, 98, 119, 143, 190; Chambliss on, 66; Covington on, 109; for cruise, 115; Hardy suggesting, 48–49; McKerley suggesting, 125–26; picture book differentiated from, 36; Salter on, 158
Social Story Creator & Library App, 51
Sokol, Michael, 5–6, 19, 94, 107, 161; on dining, 130; distraction

emphasized by, 196; on fast passes, 165; insurance discussed by, 206; JetBlue suggested by, 67; on plan, 56, 60; preparation discussed by, 47

Solomon R. Guggenheim Museum, 179

Sometimes Travel with Kids Sucks (Bigley), 13

Sonoma Raceway, 265

Southern California Railway Museum, 264

Space Center Houston, 183

Spam Museum, 270

Special Needs Parenting (Mshar), 50

special needs program, 115–16

SpecialNeedsTravelMom.com, 242

Special Service Request (SSR), 70

spectrum. *See* autism spectrum disorder

Speid, Marian, 65, 68, 73, 96, 108; on distraction, 79; on flight, 69; go-to bag recommended by, 71, 94, 111; of Marian Speid Dream Vacations, 287; on packing, 94

Spinners Pinball Arcade, 175

Splish Splash Water Park, 178

Sports Illustrated (magazine), 230

Springhill Suites by Marriott Orlando at SeaWorld, 134

SSR. *See* Special Service Request

staff, 114, 116–17, 195

STARS. *See* Steamboat Adaptive Recreational Sports

staycation, 41

Steamboat Adaptive Recreational Sports (STARS), 236

Steinman, Jake, 219–20

stimming, 60

stimuli, 43, 123

Story Land Theme Park, 177

St. Pierre, Michelle, 18–19, 78, 100, 196, 217, 287–88; itineraries considered by, 225–26

stress, 44

stroller, 80, 160, 166, 191, 286

The Strong International Center for the History of Electronic Games, 251

The Strong National Museum of Play, 179

structure, 149–50

Stull, Belinda S., 43, 73, 79, 287; on accommodation, 128; on car, 90; on dining, 112; noise discussed by, 196–97; Travel Leaders/All About Travel, LLC, owned by, 287

Subway Sleuths, 164

Sundial Special Vacations, 213

sunflower lanyards, 104–5, 148

suppliers, 217, 220–21

Surfside Beach, South Carolina, 228

Sutter, Sally J., 44, 47, 55, 73, 101, 288

swimming, 242

The System Source Computer Museum, 250

Tanque Verde Ranch, 141

The Tech Interactive, 250

technology, 16–17

Tekin, Meredith, 24, 124–25, 206

Telluride Adaptive Sports, 237

tent, 145–46, 149

Texas, Houston, 229

theme park, 165

Theme Park Nannies, 160

Thibault, Nicole, 29–30, 45, 86, 164–65, 219, 288–89

Thornton Quarry, 257

A Ticket Around the World (Diaz and Owens), 36

"Tips for Traveling with Challenging Children" (Arky, B.), 50

toileting, 211, 218

Tom, Kiana, 119

Tourette Association of America, 214

Tourette Syndrome, 213

Tourism and Autism (Hamed), 158

tours, 57–58; of airplane, 67; Bertuccio on, 187; Dolcine on, 162; Haraty on, 161; private, 161; sightseeing on, 191. *See also specific tours*

Tradewinds Island Resorts, 134

trains, 247; ASD fascinated with, 100, 103–6, 163–64; Hardy discussing,

246; Johnson, C., on, 90, 100; noise on, 102; Romans on, 103
train station, 101, 106
transfers, during travel, 93
transportation, 260–66
Transportation Security Administration (TSA), 67
travel: ADHD impacting, 9, 29; ASD impacting, 5–6, 13–21, 29, 50–51, 59–60, 142–43; benefit of, 17, 19; with children, 3–4, 7–12; culture introduced with, 53; domestic, 10–11; by family, 3–6; internationality complicating, 54; local, 30; Ludwig on, 189–90; mental health organizations relating to, 277; parent fearing, 3; picture book on, 35–36; plan for, 7; preparation required for, 92–93; rehearsal for, 158; routine disrupted by, 5, 7, 37; school influenced by, 12. *See also* cruise; trains
TravelAbility, 219–21
TravelAbility Insider (newsletter), 220
TravelAble, 284
"Traveling with Children on the Autism Spectrum" (Antell), 67–68
Travel Leaders/All About Travel, LLC, 287
Trendy Travels by Nicole, 289
trial run, 109
trigger, 200; Bertuccio discussing, 82, 129, 160; crowds as, 64; with disorders, 214; Haraty on, 67, 99, 145–46, 165; Hardy on, 216
Trips, Inc., 213
TripsRus, 213
TSA. *See* Transportation Security Administration
TSA Cares, 69, 70, 75–77, 86, 210
TSA PreCheck, 73, 75–77, 86, 210
Turkey Hill Experience, 181
Twenty-One Senses.org, 169
2019 Guide to Camping, 147, 150–51
Two Top Mountain Adaptive Sports Foundation, 240

Unique Family Adventures, 290
University of Wyoming Geological Museum, 254
UNO Parks Vilnius, 185
UPS, 47
U.S. Department of Education, 188
U.S. National Videogame Museum, 251
Utah Motorsports Campus, 266

vacation: Adlowitz on, 86–87; with ASD, 4, 54; children excited by, 42; golf included in, 230–34; around interests, 245–71; neurodiversity on, 86–87; passion building, 225; preparation for, 154; rentals for, 20; Woodbury discussing, 203
Vacation Your World, 278–79
Vail, Colorado, 232
Vail Resorts, 237
ValleyFair Amusement Park, 176
venues, 164–66, 168–85, 190–91, 198, 247. *See also* specific venues
Verified Travel Advisors (VTAs), 215–16
Vermont Adaptive Ski and Sports, 241
videogames, 250–52
videos, 61
VillaKey Villa and Condo Rentals, 134
VirtualMuseumofGeology.com, 257
Visalia, California, 220, 228
Voices from the Outside (documentary), 231
Volo Museum, 263
VTAs. *See* Verified Travel Advisors

waiting, 167, 199–201
Wallace, Nicole Graham, 42, 56, 92, 102, 149, 209; restaurant considered by, 196; with Trendy Travels by Nicole, 289
The Walters Art Museum, 175
Wasatch Adaptive Sports at Snowbird Ski Resort, 240
water, 243, 246
waterparks, 164–66
Waterville Valley Adaptive Sports, 239

Water World Colorado, 171
Watkins Glen International Raceway, 265
weather, 216, 222
WeatherTech Raceway Laguna Seca, 265
Wharton Esherick Museum, 268
wheelchair, 36, 102, 246; stroller as, 80, 160, 166, 191, 286
Wherever You Go (Zietlow Miller), 35
White Stallion Ranch, 141
Whitney Museum of American Art, 179
Wigwam Holidays, 147
Wilderness Inquiry, 188–89, 191
Wilderness Resort Hotel and Waterparks, 185
The Willard House and Clock Museum, 249
Wings for Autism, 49, 66–67, 86
Winter, Cathy, 81, 84, 208–13, 289
Wintergreen Adaptive Sports, 241
Wlodarski, Starr, 55, 77, 91, 107, 216, 290; on plan, 158
Wonder, Yvonne, 203–4
Wonder Universe, 184
Woodbury, Tara, 42, 47, 49, 55, 70–71, 84, 108, 166; on ADA, 209; on behavior, 79–80; CARD discussed by, 19, 127; on crowds, 162; cruise recommended by, 42, 112; dining considered by, 193; on distraction,

84, 103; of Escape into Travel, 290; on gate agent, 78; on restaurant, 197; rights considered by, 209; on safety, 72; on security, 73–74; on sensory issues, 160; staff alerted by, 195; on transfer, 93; on TSA Cares, 70–71; vacation discussed by, 203
Woods, Tiger, 230
Worlds of Fun, 177
WOW! Children's Museum, 171
Wurts, Topher, 163
Wyatt, Christoper Scott, 47
Wyndham Garden Hotel, 134
Wyoming Dinosaur Center, 254
Wyoming State Museum, 254

Yale University Art Gallery, 267
Yellowstone National Park, 169
YMCA of the Rockies Snow Mountain Ranch, 141
YourAutismGuide.com, 147, 150–51

Zeihr, Michelle, 29, 37–38, 80–81, 94; on accommodation, 130; on dining, 79; on elopement, 72; novelty suggested by, 149; on online review, 42; pickiness discussed by, 129; restaurant considered by, 194; sightseeing discussed by, 161–62
Zietlow Miller, Pat, 35
Zizak, Angela, 147
zoo, 30, 33–34

ABOUT THE AUTHOR

Dawn M. Barclay is an award-winning author who has spent a career working in different aspects of the travel industry. She started as an agent with her parents' firms, Barclay Travel Ltd, and Barclay International Group Short-Term Apartment Rentals (an early precursor to companies like Airbnb and HomeAway), and then branched out into travel trade reporting with senior or contributing editor positions at *Travel Agent Magazine*, *Travel Life*, *Travel Market Report*, and most recently, *Insider Travel Report*. The mother of two and a resident of New York's scenic Hudson Valley, she also writes fiction as D.M. Barr and holds leadership roles in several writer organizations.

CPSIA information can be obtained
at www.ICGtesting.com
Printed in the USA
BVHW082050150522
636469BV00001B/1

9 781538 168660